Environmental Design

An introduction for architects and engineers

Edited by

Randall Thomas

Max Fordham & Partners

E & FN SPON
An Imprint of Chapman & Hall
London · Weinheim · New York · Tokyo · Melbourne · Madras

Published by E & FN Spon, an imprint of Chapman & Hall, 2–6 Boundary Row, London SE1 8HN, UK

Chapman & Hall, 2–6 Boundary Row, London SE1 8HN, UK

Chapman & Hall GmbH, Pappelallee 3, 69469 Weinheim, Germany

Chapman & Hall USA, 115 Fifth Avenue, New York, NY 10003, USA

Chapman & Hall Japan, ITP-Japan, Kyowa Building, 3F, 2-2-1 Hirakawacho, Chiyoda-ku, Tokyo 102, Japan

Chapman & Hall Australia, 102 Dodds Street, South Melbourne, Victoria 3205, Australia

Chapman & Hall India, R. Seshadri, 32 Second Main Road, CIT East, Madras 600 035, India

First edition 1996
Reprinted 1996, 1997

Typeset in 10/12 Ehrhardt by Acorn Bookwork
Printed in Great Britain by The Alden Press, Oxford

ISBN 0 419 19930 6

A catalogue record for this book is available from the British Library

Library of Congress Catalog Card Number: 95-71097

∞ Printed on permanent acid-free text paper, manufactured in accordance with ANSI/NISO Z39.48-1992 (Permanence of Paper).

WITHDRAWN

CONTENTS

CONTRIBUTORS

Edith E. Blennerhassett, BE, MIEI, Part-time Lecturer, Kingston University and the Architectural Association, Former Partner, Max Fordham Associates

Mike Entwisle, MA, PhD, Partner, Max Fordham & Partners

Max Fordham, OBE, MA, FEng, FCIBSE, MConsE, Hon FRIBA, Visiting Professor in Building Environmental Design, University of Bath, Senior Partner, Max Fordham & Partners

Ramiro Godoy, MSc, DipIng, CEng, MCIBSE, Partner, Max Fordham Associates

Colin Hamilton, BSc, Partner, Max Fordham & Partners

Richard J. Quincey, MEng, Partner, Max Fordham Associates

Bart Stevens, BSc, CEng, MCIBSE, Partner, Max Fordham Associates

Randall Thomas, PhD, EurIng, CEng, FCIBSC, MASHRAE, Visiting Professor in Architectural Science, Kingston University, Senior Partner, Max Fordham & Partners

Bill Watts, MSc, Senior Partner, Max Fordham & Partners

FOREWORD

You can look at the making of workplaces in two very different ways. The first is to make buildings that are air-conditioned, have the deepest possible (or bearable) plans and are sealed, smooth and roughly as long as they are wide. An elegant classic of this first kind is the Willis, Faber & Dumas building in Ipswich in Suffolk.

The second way is to allow yourself to open the Pandora's box labelled 'Environmental Design'. As soon as the seal on the box is broken you find yourself tossed about in a welter of considerations that fundamentally alter the form and the aesthetic of buildings. The smooth, sealed, minimalist building box becomes impossible; the overall shape of the building must cause ventilating air to move through it, façades must open and close and shelter and protect; elevations that face into the Sun and away from the Sun may be different; materials needing small amounts of energy for their production might be used; heavy mass for energy retention, high levels of insulation and ventilating rain skins may be appropriate; and so the list would go on.

It is hard to see how such considerations might be incorporated without a change in the way that we think about architecture. It is hard to see how to make that change without thinking about the composition of architecture in a more expressive way than is common today. If we are to make this move towards responsive expressionism we will need all the help we can get from clear thinking, cool headed environmental designers. It is here that this book *Environmental Design* has a most important part to play. It is clear, logical, well illustrated and good to read; and it has the great quality of all profound work – it is easy to understand. Now let us use it to help us with our architecture.

Edward Cullinan
London, September 1995

PREFACE

In our practice we believe in a stimulating architecture that provides for the long-term needs of humanity – health and comfort – which the planet can sustainably provide. As environmental engineers, we try to ensure that the functional performance of the building and its servicing systems contribute to these ends.

In the following chapters we have tried to provide a primer and guide to the environmental principles of architecture and engineering. The whole practice has contributed by ongoing and often spirited discussion and by working together on the projects cited and many others. We owe a debt to the authors who contributed the individual chapters (cited in the table of contents) and to Charles Parrack who helped with Chapter 6. We would also like to thank the many architects whose ideas have helped to form our own.

Max Fordham

APOLOGIES AND ACKNOWLEDGEMENTS

We owe an apology to Matisse for the woman in robes in Figure 2.1 and to Rembrandt for the elephant of Figure 4.9. A number of building outlines are loosely based on projects, by Le Corbusier and Philip Johnson, that the editor admires.

We would like to thank Christine Trinder and Hannah Fulford for their exceptional patience in preparing the manuscript, Charles Parrack for his help with Chapter 6, Kitty Lux for her administrative assistance and Tony Leitch for his work on the illustrations. Colin Rice kindly read and commented on the original text. Caroline Mallinder, Lynne Maddock and Regina McNulty of E&FN Spon have been understanding, enthusiastic and friendly throughout.

Kingston University has provided stimulation and assistance for the book.

The Building Centre Trust has been most generous in its support.

The students we have taught in schools of architecture and engineering throughout the UK have helped refine our ideas.

We cannot thank our clients and the architects with whom we work too much. Without them the book would never have been published.

NOTE TO READERS

One intention of this publication is to provide an overview for those involved in building and building services design and for students of these disciplines. It is not intended to be exhaustive or definitive and it will be necessary for users of the information to exercise their own professional judgement when deciding whether to abide by or depart from it.

It cannot be guaranteed that any of the material in the book is appropriate to a particular use. Readers are advised to consult all current Building Regulations, British Standards or other applicable guidelines, Health and Safety codes and so forth, as well as up-to-date information on all materials and products.

UNITS AND ABBREVIATIONS

1. Physics and units

The SI (Système Internationale) unit of force is the newton and the unit of work (force times distance) is the newton-metre, also defined as a joule. Work and energy have the same units. Power is the amount of energy expended (or work done) per unit time – one joule per second is a watt. If a 100 watt bulb is left on for one hour the energy consumption is 100 watt-hours. This in turn can be expressed as 360 000 J (since one watt-hour equals 3600 J). Heat is a form of energy and has the same units.

Pressure is the force acting per unit area. One pascal (Pa) is a force of one newton (N) per square metre (m^2).

The unit of thermodynamic temperature in the SI system is the kelvin (K). For this reason derived units such as thermal conductivity are expressed as watts per metre kelvin (W/mK). However, the Celsius (°C) temperature scale is also in common use (the Celsius scale is also known as the centigrade scale). Absolute temperature in degrees kelvin is found by adding 273 to degrees Celsius. Thus,

$$30\,°C + 273 = 303\ K$$

The light emitted by a source (or received by a surface) is the luminous flux. The SI unit of luminous flux is the lumen. Illuminance is the luminous flux incident per unit area. One lumen per square metre is one lux.

2. Conversion factors

Length
1 micron = 1×10^{-6} m
1 m = 3.281 ft

Area
1 m^2 = 10.76 ft^2

Volume
1 m^3 = 35.31 ft^3

Mass
1 kg = 2.205 lb

Force
1 N ≅ 0.1 kg (force)
1 N ≅ 0.22 lb (force)

Pressure
1 Pa = 0.004 in H_2O
1 kPa = 0.145 psi (lb/in^2)
1 bar = 100 000 Pa

Energy, work, heat
1 kJ = 0.948 Btu
1 MJ = 0.278 kWh
1 GJ = 278 kWh
1 therm = 105.5 MJ

Power
1 kW = 1.341 hp

Thermal conductivity
1 W/m K = 0.578 Btu/(ft h °F)

Heat transfer coefficient
1 W/m^2 K = 0.176 Btu/(ft^2 h °F)

Temperature
K = 273 + °C
°C = (5/9)(°F – 32)

Temperature intervals
1 °C = 1.8 °F
1 °C = 1 K

3. Abbreviations

AC = alternating current
AHU = air-handling unit
BMS = building management system
BRE = Building Research Establishment
ca = circa; approximately
CIBSE = Chartered Institute of Building Services Engineers
db = dry bulb
HWS = hot water service
M = million; mega (i.e. 1 x 10^6)
nm = nanometres
r.h. = relative humidity
rev/min = revolutions per minute
SI = Système Internationale
wb = wet bulb
yr = year

4. Further reading

Duncan, T. (1994) *Advanced Physics: Materials and Mechanics* (4th edition), John Murray, London.

Part One

Strategies

CHAPTER 1

1.1 Introduction

Environmental design is not new. The cold environment of 350 000 years ago led our European ancestors to build shelters under limestone cliffs (Figure 1.1). More recently, English cob cottages (Figure 1.2) and Doha homes (Figure 1.3), both built of earth, demonstrate vernacular responses to light and heat.

The cob cottage evolved to provide sufficient light (say about 100 lux) under overcast skies and to limit heat loss in the winter. The Doha home evolved to provide about the same light level in bright sunlight while protecting the interior from extreme heat. It is no coincidence that the two buildings developed to give the same light level.

The rise of science in the Renaissance led to the Industrial Revolution which has enabled environmental engineers to produce reasonably comfortable conditions in almost any building in almost any climate. Some of the most visually powerful architecture of our era has taken technology and pushed it to the limits of its capabilities. The engineering systems associated with this architecture, however, have required high-grade energy to deal with the environmental problems resulting from the building design. What we need to do now is to reduce a building's reliance on fossil fuel-derived high-grade energy yet still provide comfort inside for the occupants.

What is important in achieving this? Solar energy drives the processes we live by – photosynthesis, the carbon cycle, weather, the water cycle. It is the source of our fossil fuels and it sustains the average world temperature at about 15 °C.

1.1 Reconstruction of an Acheulean hut in France.[1]

1.2 English cob cottage.[2]

The light from the Sun can replace the high-grade energy used in electric lighting. One watt of natural light more than replaces three watts of primary energy used by a fluorescent light and even more if replacing wasteful tungsten light bulbs. Wind is invaluable in providing fresh air and in helping to lower summertime temperatures.

Our activities as a species are likely to overload our habitat. We plunder fuel reserves and convert them to carbon dioxide as we generate the energy for our immediate use. Our other chemical activities can pollute the environment, as we have seen with the depletion of ozone in the stratosphere. We must change to designing our buildings to reduce our impact on the global environment.

In general, buildings are like many animals. In cold weather a source of energy is needed to keep them up to temperature and a strategy is required to prevent the heat inside from escaping. In bears, this is accomplished by a good coat of fur – in buildings it may be by insulation. The heat loss associated with ventilation also requires regulation and much recent work deals with better seals and control techniques.

In hot weather, when the external temperature is high, too much heat may enter the space. If this heat can be absorbed by the fabric of the building, the peak air temperature during the day will be less. If night-time ventilation is possible, the heat absorbed by the fabric of the building can be lost at night when the temperatures are lower. But if the buildings are lightweight and sealed, they are likely to overheat and a need for air-conditioning will result.

1.3 Doha home.[3]

What do we need to do now?

Recently, the most significant shift in thinking is to consider the building as a whole. From this perspective we should examine how the site, form, materials and structure can be used to reduce energy consumption but maintain comfort. The importance of daylight has also become more clear. Natural light is our most important and rewarding use of solar energy. Building design should aim to provide enough light whenever the Sun is above the horizon. Of course, the dangers of glare and overheating must be avoided; therefore façade design and ventilation are key elements to a successful strategy. As part of this we need to develop economic triple-glazed windows with automatically (or easily) operated blinds to control solar radiation during the day. These blinds should be thermally insulated (or they might even be separate shutters) to reduce heat loss at night. With regard to ventilation, air quality and noise will be major design factors which, in critical situations, may lead us back to fans and simple mechanical systems.

The following chapters of this book deal with these and other issues that will help us manage the transition to comfortable, healthy, well-designed, energy efficient buildings. It is based on the leading edge of current practice but it will be evident to the reader that more buildings, more monitoring and more data are needed. The book gives pointers but it is no substitute for thought – thinking about buildings has a long way to go.

Notes and references

1. Musée Régional de Préhistoire, Orgnac l'Aven, France.
2. Clifton-Taylor, A. (1972) *The Pattern of English Building*, Faber & Faber, London.
3. Illustration and photograph by Max Fordham.

Comfort, health and environmental physics

CHAPTER 2

2.1 Introduction

This chapter discusses human comfort. Part of the background for the topic is an introduction to basic scientific principles of heat transfer, the electromagnetic spectrum, light and sound and their relationship to building design. It concludes with an overview of air quality, both outdoor and indoor, and moisture issues in buildings.

2.2 Comfort and control

As an example, let us consider our own bodies, which are controlled to maintain a core temperature close to 37 °C. We function best at this temperature and variations on either side are detrimental. This temperature has evolved over a very long period under the influence of many variables.

A major factor is the need to get rid of the heat we generate as a by-product of our metabolic systems. The heat we produce varies from about 100 W at rest to about 1000 W when physically very active. A seated adult male indoors in normal conditions produces about 115 W – about 90 W of which is sensible heat and the remaining 25 W is latent heat. Sensible heat is that which we can 'sense' or feel; it is detectable through changes in temperature. Latent heat is the heat taken up or released at a fixed temperature during a change of phase, e.g. from a liquid to a gas.

Heat loss from the body occurs in several ways. Sensible heat loss from the skin or outer clothing surface occurs by convection and radiation, and there is a sensible heat loss during respiration. Latent heat loss occurs through the evaporation of moisture diffused through the skin, and of sweat, and the evaporation of moisture during respiration. The rate of heat loss from the body will depend on the air temperature, the mean radiant temperature of the surroundings (for example, in a room this is the mean of the temperatures of the walls, glazing, ceiling and floor), the air speed and the clothing worn. In temperate climates, the atmospheric water vapour pressure (i.e. the pressure exerted by the water vapour component of the air) has a slight effect on heat flow from the body;[1] in hot, humid situations the effects can be much more significant.

The naked body, if shaded from the Sun, can be quite comfortable at around 28–30 °C and at moderate relative humidities (see Appendix A for a definition) of, say, 50%. As the ambient temperature rises the body's response is (a) to direct more blood to the surface, which increases the skin temperature and heat loss, and (b) to sweat to lose heat through evaporation. We begin to feel uncomfortable when these responses become significant.

When the ambient temperature drops, the body will limit heat loss by reducing blood flow to the surface, which reduces the skin's temperature, and by not sweating. Goose pimples are caused by tiny muscles lifting hairs on the skin, which will decrease the air flow and heat loss across the surface. This mechanism was more effective in our more hairy ancestors, and one way we have compensated for cold has been through the use of clothing. With air temperatures between, say, 20 and 26 °C we limit our heat loss with clothing but generally feel comfortable. The amount of clothing we require increases as the temperature decreases and the general atmosphere becomes less comfortable. If the heat loss through our clothing becomes too great, we generate more heat specifically for temperature control by shivering.

Two points here are relevant to building design. Firstly, to make buildings comfortable, they should be kept within a suitable temperature range which is not as wide as that in an uncontrolled external environment. Secondly, our bodies are capable of maintaining a very stable core temperature with a fairly constant metabolic heat output over a wide range of external temperatures. This is done, with little or no additional energy expenditure, by a combination of control processes including sweating, altering the blood flow (and therefore the heat loss to the skin) and changing clothes to suit conditions.

Modern buildings have achieved the first objective of maintaining fairly constant internal conditions to comfort standards with the use of significant amounts of energy to provide heating or cooling to compensate for the changing external environment. The amount of energy used could be reduced significantly if buildings adopted the principles of animal physiological control. To do this one first needs an understanding of human comfort and how energy is expended to provide it.

Comfort is a subjective matter and will vary with individuals. It involves a large number of variables, some of which are physical with a physiological basis for understanding. Classically, for thermal comfort they include:

- air temperature and temperature gradients
- radiant temperature
- air movement
- ambient water vapour pressure
- amount of clothing worn by the occupants
- occupants' level of activity.

Other factors influencing general comfort are light levels, the amount of noise and the presence of odours. Individuals are also affected by such psychological factors as having a pleasant view, having some control of their environment and having interesting work. For some variables it is possible to define acceptable ranges but the optimal value for these will depend on how they interact with each other, e.g. temperature and air speed, and personal preference.

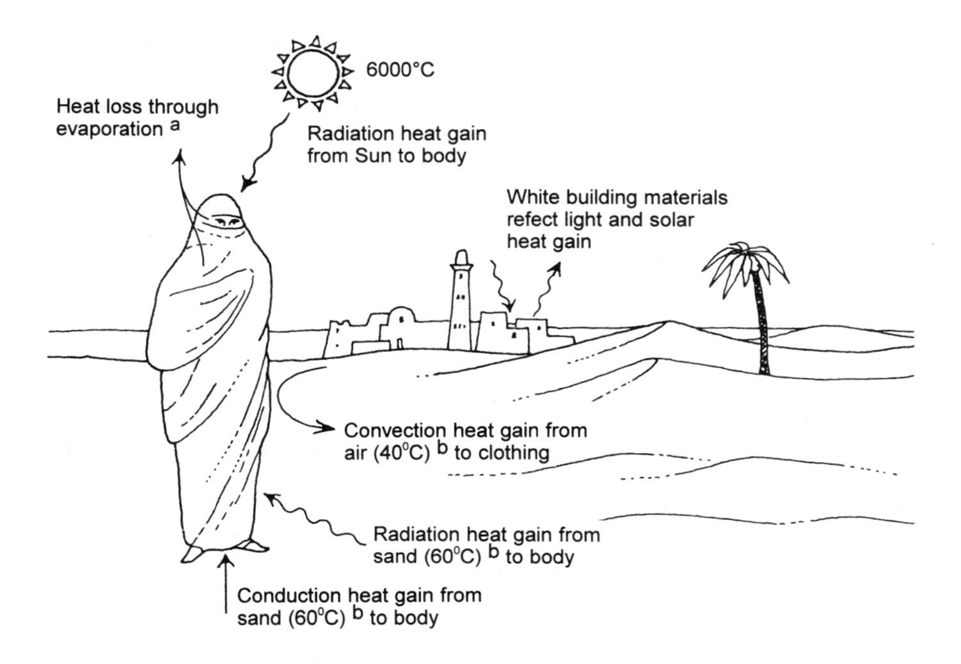

[a]The loose-fitting robes touch the skin at the shoulders only. Depending on the exact air and body temperatures an upward draught of air can help keep the wearer cool by increasing the rate at which sweat evaporates.[2] This thermally induced upward draught is known as stack-effect ventilation and is also common in buildings (Chapter 9).
[b]Approximate temperatures.

2.1(a) Heat transfer mechanisms: the Bedouin by day.

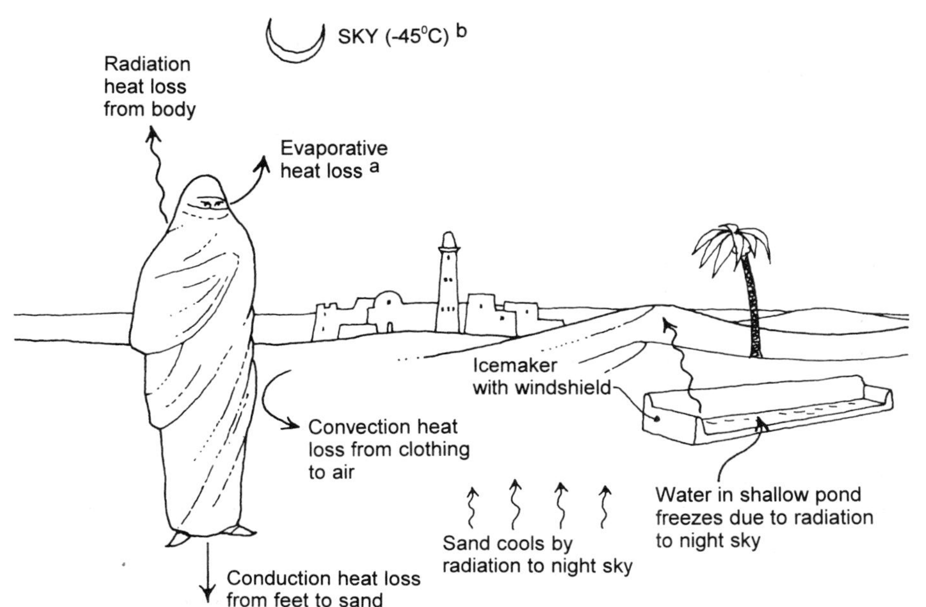

[a]Evaporative heat losses occur from the skin and respiration. At night clothing is used, especially for its insulating effect.
[b]The night sky has an effective temperature of –45 °C for radiation from the Earth.[3]

2.1(b) Heat transfer mechanisms: the Bedouin at night.

2.3 Thermal comfort and heat transfer

To be thermally comfortable one must not feel too hot or too cold, or have any part of the body too hot or too cold. The physiological basis for this is that the amount of heat being produced by the body is in balance with the heat loss, comfortably within the body's control mechanism. Furthermore, there should be no parts of the body having to operate outside the comfortable limits of the control systems (such as a cold neck from a draught or a hot face from an open fire).

There are several mechanisms which transfer heat and therefore affect this balance. Heat always flows from hot bodies to cold ones. Individuals may be gaining or losing heat depending on the relative temperatures of their bodies and their surroundings. They may indeed be gaining heat through one mechanism and simultaneously losing it through another.

As an illustration, on a very hot day in the desert the air and sand are both likely to be hotter than a person's body. The body is therefore heated by the sand on which one stands, by the air and by the Sun. Depending on the exact air and body temperatures, the only cooling mechanism may be by evaporation through respiration and sweating. Figure 2.1 illustrates the four heat transfer processes of conduction, convection, radiation and evaporation/condensation. These basic physical processes apply both to humans and buildings, and we shall examine them in turn.

Conductive heat transfer

In this process heat travels through matter by one hot vibrating molecule shaking the cooler ones adjacent to it, thereby making them hotter and passing the heat along (or, more technically, by the transfer of kinetic energy between particles). Metals tend to be good conductors and so have high thermal conductivities. (The thermal conductivity of a material is the amount of heat transfer per unit of thickness for a given temperature difference. Organic materials such as wood and

Table 2.1 Properties of selected materials [4,5]

Material	*Density (kg/m^3)*	*Thermal conductivity (W/m K)*	*Specific heat capacity (J/kg K)*
Bricks	1700	0.73[a]	800
Concrete, dense	2000	1.13	1000
Glass fibre quilt	25	0.035	1000
Asphalt	1700	0.50	1000
Aluminium	2700	214	920
Water (20 °C)	1000	0.60	4187
Sand (dry)	1500	0.30	800

[a]Mean of internal and external brick types. Consult manufacturers' data for precise values.

plastic tend to be poor conductors. Aerated materials, which have solid conduction paths broken by air or gas gaps (foam, glass fibre quilt or feathers) are very poor conductors but are good insulators as they have low thermal conductivities.) Table 2.1 shows a range of thermal conductivities and other properties of some materials.

Crudely, if an object is cold to touch it is probably a good conductor as it conducts the heat away from your hand. If it is warm it is a good insulator as the heat is not drawn from your hand. This, of course, assumes that the object is colder than your skin temperature. Hotter conducting materials will feel hotter than insulating ones as more heat will be conducted to your hand.

One of the pitfalls of this test illustrates another property of materials: the capacity to store heat. At room temperature aluminium foil may feel cold for a short period, but then warm. This is because the small mass involved heats up quickly and not because it is a bad conductor. A thick aluminium saucepan would feel cold as there is adequate mass to absorb a reasonable quantity of heat. Similarly, one may risk holding a loose piece of foil straight from the oven with bare hands, but not a solid baking dish.

The amount of heat a material can store, or its thermal mass, is the product of multiplying its mass, its specific heat capacity and the increase in temperature. The specific heat capacity (Table 2.1) is the amount of heat that a material will store per unit of mass and per unit of temperature change. Note the comparatively high specific heat capacity of water.

Convective heat transfer

Convective heat transfer is the process by which heat is transferred by movement of a heated fluid such as air or water. If we consider a hot surface and a cold fluid, the fluid in immediate contact with the surface is heated by conduction. It thus becomes less dense and rises, resulting in what are known as natural convection circulation currents. Convection that results from processes other than the variation of density with temperature is known as forced convection and includes the movement of air caused by fans.

Radiant heat transfer

In conduction and convection, heat transfer takes place through matter. For radiant heat transfer, there is a change in energy form, and bodies exchange heat with surrounding surfaces by electromagnetic radiation such as infrared radiation and light (Figure 2.2). Surfaces emit radiated heat to, and absorb it from, surfaces that surround them.

The amount of heat emitted from a surface depends on its emittance and its temperature; as the temperature increases the heat emitted increases (it is, in fact, proportional to the fourth power of the absolute temperature). The emittance itself is a function of the material, the condition of its surface, wavelength and the temperature. The amount of radiation absorbed by a surface is its absorptance. The absorptance depends on the nature of the surface, the spectral distribution of the incident radiation and its directional distribution. (In contrast

Table 2.2 Emittances and absorptances of selected materials [6]

Item	Emittance (at 10–40 °C)	Absorptance (for solar radiation)
1. Black non-metallic surfaces such as asphalt, carbon, slate, paint	0.90–0.98	0.85–0.98
2. Red brick and tile, concrete and stone, rusty steel and iron, dark paints (red, brown, green, etc.)	0.85–0.95	0.65–0.80
3. Yellow and buff brick and stone, firebrick, fireclay	0.85–0.95	0.50–0.70
4. White or light cream brick, tile, paint or paper, plaster, whitewash	0.85–0.95	0.30–0.50
5. Bright aluminium paint; gilt or bronze paint	0.40–0.60	0.30–0.50
6. Polished brass, copper, monel metal	0.02–0.05	0.30–0.50

with emittance, the temperature of a surface has only a very small effect on its absorptance.) Table 2.2 gives some data on emittances and absorptances.

As can be seen, most surfaces commonly used in buildings (paint, dull metal, glass, etc.) have high emittances (about 0.8–1). Black non-metallic surfaces can be seen to have both high emittances and absorptances. Shiny metallic surfaces have both low emittances and absorptances.

When radiation strikes a surface it is absorbed, reflected or transmitted (i.e. it passes through the material struck) with the relative proportions depending on the characteristics of the surface and the wavelength of the incoming radiation. Absorbed radiation will, of course, cause a material to heat up.

The varying nature of materials can be exploited – for example, heat loss through windows can be reduced by using a low emissivity coating on the glass; solar collectors are usually matt black to maximize their heat absorption.

For people, the mean radiant temperature of their surroundings is important for comfort, as is the variation or uniformity of radiant temperature – imbalances that make one hot and cold on different sides can be disagreeable. Radiant heat transfer can be felt most obviously, for example, by standing outside facing a bonfire on a clear night. One's face will be hot from the radiation of the fire and one's back (if lightly clad) will feel colder, partly because of convective heat transfer to the cool night air and partly from body heat being radiated to the surrounding night sky (Figure 2.1(b)).

Evaporative heat transfer

Molecules in a vapour state contain much more energy than the same molecules in a liquid state. Thus, energy must be added to turn a liquid into a gas. The amount of heat required to change liquid water into a vapour is the latent heat of evaporation. This heat is removed from the liquid – which is thus cooled – and transferred to the vapour. Evaporation causes cooling of surfaces.

In condensation the process is reversed and the latent heat of evaporation is transferred from the vapour to the surface.

The amount of energy transferred in evaporation and condensation is considerable compared to that required to heat or cool a liquid or gas; for example, to vaporize 1 kg of water at boiling point it takes about 500 times as much heat as is required to increase the temperature of 1 kg of water by 1 °C. Steam at 100 °C is more likely to burn one's skin than dry air from a hot oven at over 200 °C.

The maximum amount of water vapour that can be held in a fixed mass of air is related to the temperature of the air (Appendix A). At 20 °C it can carry up to 15 g of water per kg of air; at 0 °C it can hold only 4 g of water.

The direction of vapour flow to or from a wet surface is dependent on the quantity of water in the air and the temperature of the surface. The air immediately above a wet surface is assumed to be at the same temperature as that surface, and 100% saturated. This will define a water vapour content in grams of water per kilogram of dry air. For that surface to lose heat by evaporation, the surrounding air must have a lower amount of water vapour per kilogram of dry air. In very hot and humid conditions the water content of the air is high and so the surface temperature must be comparatively higher to lose heat. The rate of evaporation is determined by the difference in water vapour content and the air speed across the surface.

Comfort levels

All the processes described above contribute to our thermal balance, which is the sum of the effects of the heat exchanged by the body with its environment. We feel comfortable if we can maintain our thermal balance without much effort, and uncomfortable in our environment if we have to shiver to generate heat or sweat profusely to lose it.

Comfort levels will obviously fall well within these limits of shivering and sweating. The physics of heat transfer would suggest that optimal conditions will depend on a person's activity and clothing. This has been confirmed by research on occupant preferences in a variety of thermal conditions.[7] In casual summer clothing – tee shirt, shorts and sandals – the optimum temperature for sedentary work at 50% r.h. is about 25–26 °C. For more formal and winter clothing – for example, suits – the optimum temperature is 20–21 °C.

This has obvious implications. Firstly, the dress code for occupants of the building has an influence on the optimum temperature of the building. In winter the heating energy is reduced if the occupants wear heavier clothes, and in summer the extent to which the building has to be cooled down can be minimized if more casual wear is allowed. Secondly, all the occupants would ideally have a similar level of clothing insulation but, in fact, there are often significant differences in the thermal characteristics of what men and women wear.

Discomfort is felt if there are large variations in the environmental conditions around the body such as:

- Wide variations in air temperature. Because warm air rises it is quite common to have a temperature gradient in a room such that it is cool near the floor and hot near the ceiling. This can give the unsatisfactory situation of a hot head and cold feet.
- Wide variations in radiant temperature. This can be felt by sitting next to a large cool window or a high-temperature source such as an open fire.
- Draughts. Draught discomfort depends on the difference between the skin and air temperature, the air speed in the room and the turbulence of the air movement;[8] (turbulent flow is contrasted with smooth or laminar flow).

Thermal comfort is the subject of an enormous literature, and a number of standards which are regularly reviewed (see, for example, Reference 9) are available to give guidance. If we try to state briefly what a designer should provide, it is important to allow occupants some control of their environment. Another aspect is to think in terms of acceptable internal temperature in relation to ambient temperatures. In the UK, which is not characterized by extremes of temperature (the heat wave of 1994 saw maximum daily temperatures in the 25–30 °C range in London and minima in the 15–22 °C range), much of the comfort research dates back to a transition period when lightweight buildings were replacing heavyweight designs and discomfort was resulting. Research in the 1960s indicated that if the peak internal temperatures could be kept at 24 °C for days that had a minimum of 11 °C, a mean of 18 °C and a maximum of 25 °C, then there would be a low level of complaints of overheating; it also found that comfort standards might be relaxed to allow temperatures to rise to 27 °C (a differential of 2 °C).[10] The same study showed the very important effect of noise on comfort, with people working in offices in quiet areas much less likely to experience discomfort at higher temperatures than those in noisy areas.

A guideline for UK schools says that during the summer the recommended design resultant temperature (this is the temperature recorded by a 100 mm diameter globe thermometer and takes into account air temperature and radiant temperature in equal proportions) measured 0.5 m above floor level should be 23 °C with a swing of not more than 4 °C about the optimum.[11] It goes on to say that it is undesirable for the resultant temperature to exceed 27 °C during normal working days over the school year, but an excess for 10 days during the summer is considered a reasonable predictive risk. The precise meaning of some of these statements is open to interpretation but they offer a broad guidance.

More recent work by the BRE suggests that summer-time peak temperatures in a 'formal' office might be 23 ± 2 °C.[12] In the author's own well-ventilated office in a fairly noisy area with good individual control possible, there were very few complaints at internal temperatures of 30 °C when the temperature outside reached 28 °C (with a slight breeze). It should also be said that we have an informal dress code and that all the occupants have good external views.

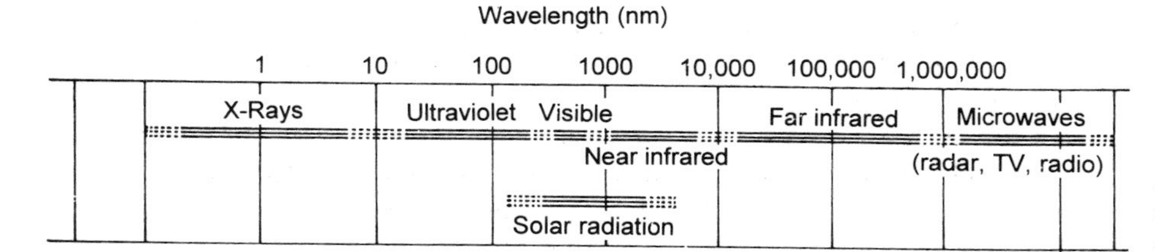

2.2 The electromagnetic spectrum.

2.4 The electromagnetic spectrum

Electromagnetic radiation is energy in the form of waves generated by oscillating magnetic and electrical fields. This radiation covers a spectrum of wavelengths, as shown in Figure 2.2.

The spectrum has no definite upper or lower limit and regions overlap; the visible region is roughly from 400 to 760 nm.

The wavelength multiplied by the frequency equals the speed of propagation, which is the speed of light. The energy of the radiation is proportional to its frequency. Intensity of radiation decreases with the square of the distance from a point source; thus, if the distance from the source is doubled, the intensity falls to one-quarter.

All matter warmer than absolute zero (0 K or –273 °C) produces a spectrum of radiation which varies with its temperature. A blackbody is given this term because it absorbs all the energy incident on it; these perfect absorbers are also perfect radiators, or emitters, of energy. (Most bodies are in fact less perfect and are treated as grey bodies.) The hotter the body, the more total energy is radiated and the higher the energy and frequency of the radiation; correspondingly, the wavelength of the emitted energy is lower (Figure 2.3). The Sun, whose spectrum is essentially that of a blackbody at 6000 °C, produces a broad range of radiation from ultraviolet through visible light to infrared. A filament light bulb produces most of its energy as invisible infrared radiation with some visible light. This is why a filament lamp is comparatively inefficient compared to a fluorescent one, which is designed to produce most of its radiation in the visible spectrum.

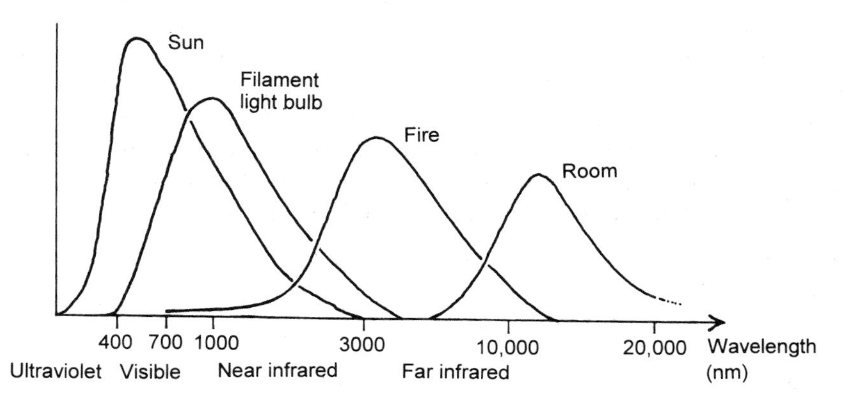

2.3 Spectra of radiation from various bodies.[13] (*N.B.* The vertical axis is not to scale and does not give relative intensity.)

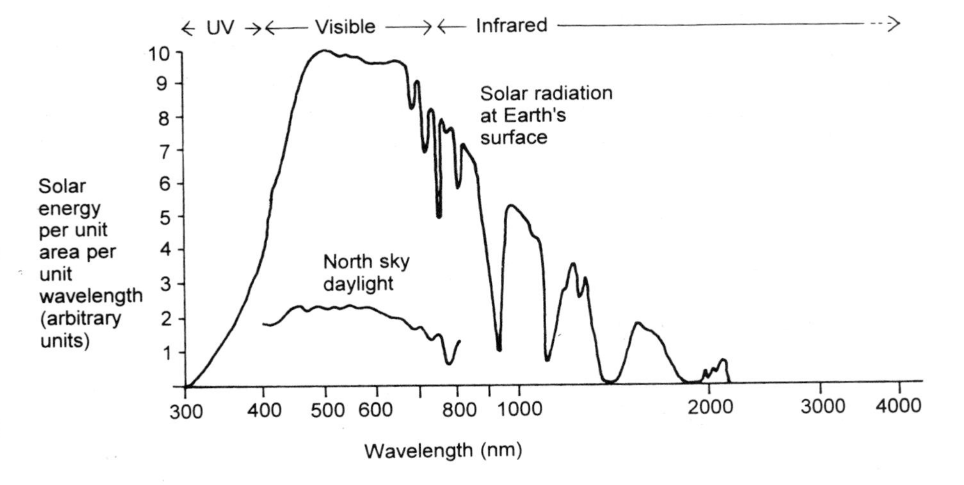

The approximate spectral composition and intensity of north sky daylight of 5700 K is also plotted for comparison (see Figure 8.7).

2.4 Spectral distribution of solar radiation at the Earth's surface.[15]

Figure 2.4 shows the spectral distribution, i.e. the amount of radiation at various wavelengths, of solar radiation at the Earth's surface. The atmosphere filters out most of the Sun's ultraviolet and much of the infrared radiation; the ultraviolet being particularly removed by the ozone layer in the upper atmosphere. At the Earth's surface, approximately 10% of the Sun's radiated energy is in the ultraviolet range of 290–400 nm, 40% in the visible range of 400–760 nm and 50% in the infrared range of 760–2200 nm.[14] (These figures vary somewhat with how the wavelength band radiation limits are defined and according to which reference one consults.)

Life on Earth has evolved to make use of incident solar radiation – human vision and photosynthesis being two examples of this. The depletion of the ozone layer has recently become an area of major concern. We and other forms of life have not developed biological mechanisms to protect us from large amounts of high-energy ultraviolet (UV) radiation as most of this is filtered out by the ozone layer. The likely consequences of continued loss of ozone in the atmosphere include a higher incidence of skin cancer and eye cataracts, and damage to land and marine vegetation by UV radiation. Chemicals, mainly chlorofluorocarbons (CFCs) and hydrochlorofluorocarbons (HCFCs), which are depleting the ozone layer, are discussed in Chapter 6; CFCs and HCFCs also contribute to global warming as discussed below and as shown in Figure 2.5.

Water vapour, carbon dioxide and ozone in the atmosphere absorb infrared radiation from the Sun and from the surface of the Earth. This insulates the planet and keeps it warm – unlike the surface of the Moon which swings greatly in temperature, varying from 100 °C on the sunlit surface to –150 °C at night. The insulating effect of water vapour in clouds can be seen when comparing a clear night to a cloudy one. On a cloudy night the ground cools down comparatively slowly as it is radiating to a thick blanket of water vapour which is absorbing the heat and radiating much of it back to Earth. On a clear night the sky has less water vapour, and thus much more of the infrared radiation from the ground escapes into space.

The Earth will, on average, lose the same amount of heat to the Universe as it gains from the Sun. If the atmosphere becomes a better insulator of infrared radiation, the Earth's surface will become warmer in order to lose the same amount of heat coming from the Sun. There is great concern that higher 'greenhouse' gas levels and, in particular, carbon dioxide from burning fossil fuels

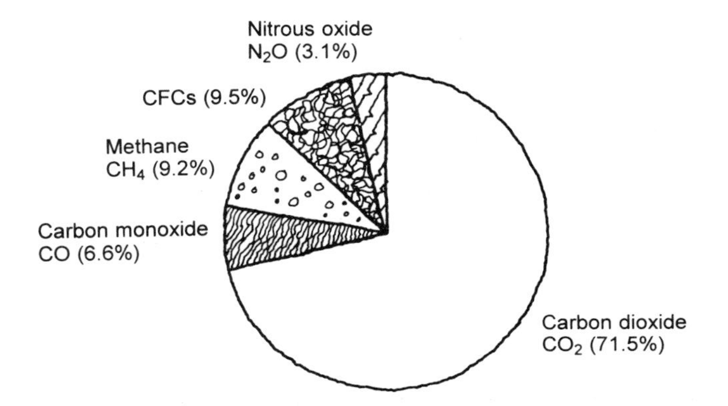

2.5 Relative contributions of gases to global warming in 1985.[16]

(Figure 2.5) are subtly increasing the insulation of the atmosphere, thus causing a rise in the world's average temperature. It is feared that a small increase in temperature will melt sufficient amounts of water currently frozen in polar ice-caps to raise the ocean's water level to flood low-lying land all over the globe.

2.5 Light

As mentioned above, light is the visible portion of the electromagnetic spectrum. Wavelengths are associated with colours, as can be seen in Figure 2.6, which also shows that the eye is most sensitive to green light at about 550 nm. White light is a mixture of various wavelengths.

Light levels outside vary enormously from 100 000 lux in bright sunlight to 0.2 lux in bright moonlight and 0.02 lux in starlight. Our eyes will register information over the range of brightest sunlight down to about 0.005 lux.

While this is our full range, the amount and content of information we can register drops off at low light levels. A young person with good eyes can probably thread a needle in 100 lux, read a theatre programme in 10 lux and distinguish large objects in 0.005 lux. It would be a strain if one was always to do these tasks at these light levels and older people would find them difficult to do at any time. For this reason the recommended light levels associated with various tasks tend to be 5 to 10 times greater than the absolute minima (Chapter 8).

Our eyes can deal with the range of lighting levels by adapting to different average ambient light levels. They cannot, however, register information across the entire range at the same time as they take time to adapt when going from a

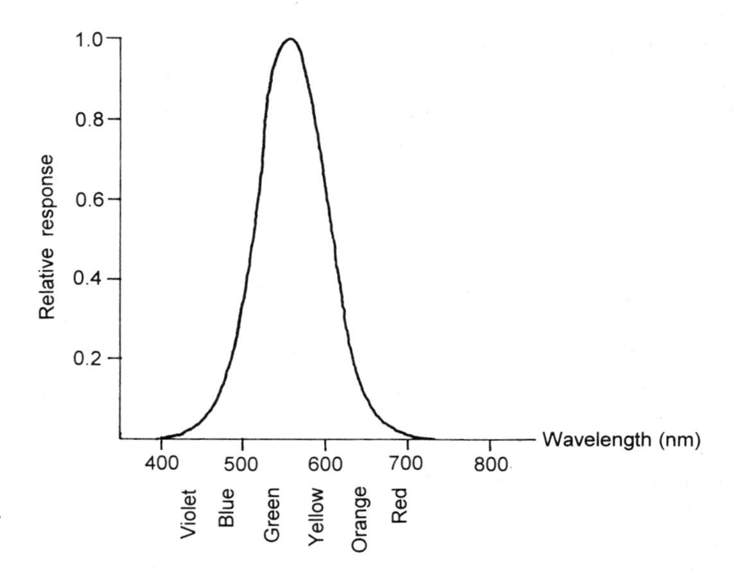

2.6 The spectral response of the average human eye for photopic vision.[17]

light space to a dark one, and vice versa. Consequently, wide ranges of brightness within one's field of view are uncomfortable as the eyes do not know which level to accommodate. If they adjust to the higher level, the information from the lower level is lost; and in adjusting to the lower level, the eye gets painfully overloaded from the higher level.

This is a problem of glare and is illustrated by the classic interrogation technique with the lights facing the victim in an otherwise darkened room. Outside, in daylight, the lights would not be much brighter than their surroundings and the effect would be lost.

Light is directional and travels in straight lines. It is reflected from surfaces: reflection from rough or irregular surfaces is diffuse. Surfaces such as mirrors reflect light directly and this is known as specular reflection.

The colour of a surface determines how much light is reflected and how much is absorbed. White or untinted mirror finishes will reflect almost all the light; black will reflect very little; and shades will vary between these two (Table 8.3). (Reflectance is the ratio of light reflected from a surface to that incident upon it.) The colour of a surface we perceive is the colour of the light hitting the surface less the colours that are absorbed by the surface itself. In white light a green leaf absorbs the blue, red and yellow colours and reflects the green. In the light from a yellow sodium street lamp (which has no green component), a leaf will absorb the yellow and reflect nothing and thus appear black.

Light can reach a surface from a source such as a light bulb or the Sun directly, indirectly by being reflected from surrounding surfaces, or by a combination of both. The colour and reflectance of surfaces do have a dramatic effect on the indirect light component, and therefore the total light level, in a space. The light level in a darkly decorated room will be much less than that in a white-painted room given the same light input.

The ratio of direct to indirect light affects the visual 'feel' and comfort of a space. All direct light with no reflected component (e.g. a matt black room with a point source of light) will show objects in very high contrasts and sharp shadows. At the other extreme a completely diffuse lighting scheme (e.g. an evenly illuminated ceiling with white walls and floor) will be shadowless and without texture.

2.6 Sound

Sound is a waveform like light and is governed by some of the same physical principles. Unlike light, sound is a pressure wave (caused by something vibrating) travelling through a solid, liquid or gaseous medium. It cannot travel through a vacuum. In air it is relatively easy to produce a noise that will travel quite far, and not necessarily in the line of sight. Production of sounds is therefore a useful means of communication that is found throughout the animal world.

The pitch of a sound depends on the frequency of the sound wave and is the equivalent of colour in light – high-pitched sounds are of high frequency. The human voice produces sound in the range of 200–2000 Hz (cycles per second). The human ear is sensitive to a range of sound frequencies from about 15 Hz, which corresponds to a very low rumble of a distant bus or the lowest organ note, up to 20 000 Hz; door squeaks and the chirp of some insects have a frequency of about 17 000 Hz. However, the ear is less sensitive to high and low frequencies than to those in the middle range.

Sound levels (loudness) are commonly measured in decibels (dBA – see Appendix C for further discussion). This is a scale which takes account of the intensity of all the audible frequencies and weights them in accordance with the ear's sensitivity. It gives a single-valued number that correlates well with the human perception of relative loudness. A sensitive human ear can detect sound down to about 10 dBA. Normal speech is conducted at about 55–70 dBA. The threshold of pain is approximately 130 dBA. The decibel (dB) is not a linear value but a logarithmic one and a huge range of about 10^{17} is covered by the sounds we hear from the quietest whisper to booster rockets. The ear adapts to different sound intensities and takes time to readjust to changes in levels.

It has been established that long-term exposure to high noise levels will cause premature hearing degradation, if not loss. The current UK Noise at Work Regulations suggest that employers need to protect employees exposed to 90 dBA;[18] EEC guidelines suggest that at 85 dBA workers need to be informed about risks to their hearing arising from exposure to noise.[19] Although, in most situations, noise levels in buildings will be well below this figure, in some cases, such as very reverberant canteens and swimming pools, higher levels can be reached.

The ear and brain are very good at filtering sound to extract information such as speech from a background of noise. There are, however, limits and the greater the noise/speech ratio the less information is received. The brain is capable of filling in the gaps in the information to a greater or lesser extent depending on one's prior knowledge of the subject and one's skills of interpolation.

This filtering of speech from noise is particularly noticeable at a party. At the start, when few people have arrived, one can talk normally as the background noise level is low. As more people arrive, the number of people speaking, and therefore the total sound level, goes up. To be heard over the background noise, people must raise their voices and so the noise level rises further. To increase the probability of their speech being understood in spite of the background noise, people get closer and closer together.

It has been found that at a background sound level of 48 dBA the maximum distance for normal speech intelligibility is about 7 m; at 53 dBA the distance falls to about 4 m and so on.[20] Raising the voice increases the distance, and teachers know this well. The Department of Education and Science make recommendations for the background noise levels that are acceptable in a classroom to maintain the teacher's intelligibility to the pupils.[21] For example, the maximum background noise level (BNL) for a large lecture room is given as 30.

Background noise also provides privacy and the lack of it can be just as unwelcome as an excess. In a completely quiet library every private conversation can be heard – some background noise will shield this and make those talking feel less conspicuous. From the opposite point of view, in quiet, open plan offices conversations can sometimes be overheard so clearly as to be a distraction to others trying to work.

Sound waves can be reflected or absorbed by surfaces depending on their construction. Generally, hard smooth finishes will reflect sound. Sound can be absorbed in a number of ways, and three common ones in buildings are dissipative (or porous) absorbers, membrane absorbers and cavity absorbers. Here we shall deal only with the first two, and briefly at that. Porous absorbers allow the pressure wave into the surface of the material. Friction between air particles and the material results in dissipation of sound energy as heat. The effectiveness of this absorption depends on the thickness of the absorptive material compared with the wavelength of the incident sound. Thicker materials will absorb longer wavelengths better than thinner ones. Membrane (or panel) absorbers first convert the energy of the pressure wave into vibrational energy in the panel facing, and then further loss occurs in the air space behind the panel. Suspended ceilings and raised floors are two common membrane absorbers. Generally, no single surface provides adequate absorption over a wide frequency range. Membrane absorbers (panels) tend to be better at lower frequencies, and porous absorbers (for example, soft furnishings such as heavy curtains) tend to be better at higher frequencies.

Sound will normally come either directly from a source or indirectly, having been reflected from the surroundings, or as a combination of the two. The area and absorption of the surfaces in a room will affect the amount of indirect sound and therefore the total sound level within it. A living room with thickly upholstered furniture, deep carpet, heavy curtains and the odd wall hanging will sound quiet or 'dead', and the volume of the TV will have to be quite high. The same room with no soft furnishings will sound more 'live' and the volume of the TV can be reduced to achieve the same sound level.

The overall measure of the absorption within a space is the reverberation time (RT), which was defined by the eminent acoustician Sabine as the time taken for a sound to decay by 60 dB after the source has stopped. The reverberation time for a space will vary with the frequency of the sound and depends on the type of absorption in the space. Just as the colour of the paint on the walls will affect the colour and amount of light in a room, the absorption will affect the tone and intensity of sound in a space.

For music, high reverberation time has the desired effect of blending discrete notes together. The same effect is less appreciated for speech. Discrete words may be blended together, making them unintelligible. There is therefore a conflict between halls used for music, which requires high reverberation times, and halls used for speech, which needs lower times for intelligibility. Normally, resolution is by designing for a compromise reverberation time between the two extremes. For example, in a room with a volume of 200–300 m^3 the optimum reverberation time for speech might be about 0.7 s and, for music, 1.3 s,[22] and a value between these two would probably be used.

The problem referred to above of conversations being overheard in quiet open plan offices is related to reverberation time. High reverberation times help to provide acoustic privacy because the overheard communication is a mixture of direct sound and garbled reverberant noise. On the other hand, in a space with highly absorbent finishes and low reverberation times there is very little reverberant noise to mask direct sound, so low levels of sound from far away are still intelligible. Thus, where this is a problem one approach is to try to make the space more reverberant. Another is to install active noise generators.

One space can be acoustically separated from another by using solid partitions and by ensuring that no direct air paths connect the two. The heavier the partition, the more difficult it is for the air pressure waves to vibrate it and the greater the separation. This relationship is characterized by the mass law (Figure 2.7). (The sound insulation of an element is basically the difference between the sound level in one room with a noise source and the sound level in an adjacent room that is separated by that element.)

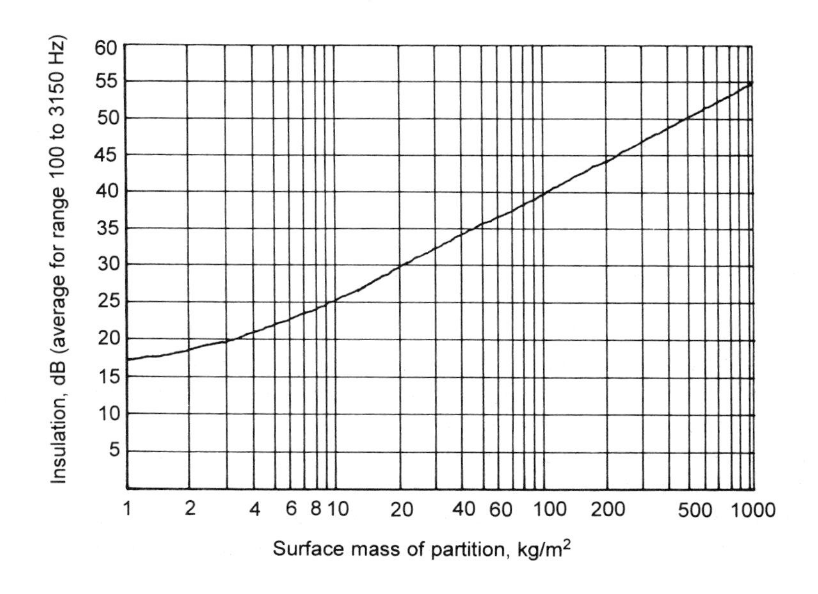

2.7 Sound insulation as a function of surface mass for single-leaf construction.[23]

Masonry walls between separate dwellings often have masses of about 380 kg/m^2 and, thus, sound insulation of about 47 or 48 dB. Any air gaps in these or other construction elements will degrade the performance.

Double-leaf walls, i.e. two skins separated by an air gap, can in many circumstances provide adequate insulation with less mass than that needed with a single-leaf wall.

Air paths required for ventilation can be designed such that there is little or no direct line of sight route for the sound and the walls can be lined with absorbent material so that there is little reflected sound. To increase the surface area for sound absorption and reduce the chance of sound passing through without being absorbed, air ducts can be divided up with 'splitters' into a series of smaller, thin-lined ducts.

Because of the importance of acoustics in buildings we shall return to these issues several times in the following chapters.

2.7 Air quality

Clean, dry air near sea level has the following approximate composition:[24]

Nitrogen (N_2)	78%
Oxygen (O_2)	21%
Argon (Ar)	1%
Carbon dioxide (CO_2)	0.03%

In addition, there are trace amounts of hydrogen, neon, krypton and a number of other gases as well as varying amounts of water vapour and small quantities of solid matter. Because of rising CO_2 production resulting from human activities, the CO_2 level in much of the atmosphere's air is higher, at about 0.035%.

Pollutants in the air include:

- Nitric oxide (NO) and nitrogen dioxide (NO_2), known collectively as NOx
- Sulphur dioxide (SO_2) and, to a lesser extent, sulphur trioxide (SO_3), known as SOx
- Volatile organic compounds (VOC), including benzene and butane
- Carbon monoxide
- Lead
- Ozone.

A number of these pollutants result from the combustion of fossil fuels to meet our energy demands.

In addition, there is a great deal of 'dust' in the air. In fact this is a mixture of dusts, smokes and fumes which are particulate matter and vapours and gases which are non-particulate.[25] As an indication of sizes, smoke particles are 0.01–0.5 micrometres (μm); mists and fogs are under 100 μm; and pollen is in a range of about 10–100 μm. Heavy industrial dust can range from 100 to 1000 μm and higher.

The amount of solid material in the atmosphere obviously varies enormously, but as a guideline in metropolitan areas it is in the range of 0.1–1.0 mg of mate-

Table 2.3 Analysis of typical atmospheric dust[27]

Range of particle size diameter (μm)	*Percentage of total mass of particles (%)*
30–10	28
10–5	52
5–3	11
3–1	6
1–0.5	2
Below 0.5	1

rial per 1 m^3 of air (mg/m^3) and in rural and suburban areas 0.05–0.5 mg/m^3.[26] Table 2.3 gives a typical analysis for average urban and suburban conditions combined.

There is growing concern about the health effects of most air pollutants and also the combined effect of dust.[28,29] PM10, the name given to particulate matter of diameter 10 μm and less, is now thought to be a cause of cardiovascular and respiratory diseases.

One of the most serious effects of air pollution is 'acid rain', caused chiefly by oxides of sulphur (SOx) and nitrogen (NOx) emitted during fossil fuel combustion and metal smelting.[30] 'Acid rain' is the formation in the atmosphere of acids which are then returned to earth resulting in acidification of streams and lakes, damage to trees and degradation of stonework in historic buildings.

Architects and engineers can help reduce external air pollution by making their buildings energy efficient, thus reducing the burning of fossil fuels, and by specifying less polluting combustion equipment. We shall return to this in Chapter 9.

Ventilation is the provision of air to a building. One reason for ventilation is that the occupants need oxygen to oxidize their food (which contains carbon and hydrogen) to produce the energy needed to live. In the process, carbon dioxide and water are formed. The ecological cycle continues with plants taking up carbon dioxide and water in photosynthesis and producing oxygen and food. Our oxygen requirements are met by about 0.03 litre of air per second per person (l/s person).[31]

Removing the carbon dioxide we produce requires higher ventilation rates than those needed to provide oxygen. The occupation exposure limit for CO_2 is 0.5%. Maintaining the room CO_2 level at 0.5% for people involved in light work requires 1.3–2.6 l/s person of fresh air (assuming 0.04% of CO_2 in fresh air); to limit the room CO_2 level at 0.25% requires 2.8–5.6 l/s person.[32]

Ventilation is also required to remove indoor pollutants, to take away moisture to reduce the risk of condensation (Chapter 4) and to take away heat in the summer to maintain comfort (Chapter 9).

Indoor pollutants, some of which originate outside and some inside, include nitrogen dioxide (NO_2), carbon monoxide (CO), carbon dioxide (CO_2), radon, formaldehyde, sulphur dioxide (SO_2), ozone (O_3), mineral fibres, tobacco smoke, body odours and hundreds of other substances. Acceptable levels for all of these do not exist and it is not easy to define the quality of air that is suitable. The answer affects health and energy conservation. Ideally, the best approach is to reduce the pollutants at source – a common example at present is to prohibit smoking in many areas. After that there are two approaches: the first is simply to reduce their concentration by diluting them with more 'fresh' air; the second is to filter out contaminants. Both require energy.

There is not yet complete agreement on how much fresh air is required in buildings. In the UK, the CIBSE recommends 8 l/s person in offices in the absence of smoking.[33] In the US an ASHRAE standard for ventilation for acceptable indoor air quality gives a figure of 10 l/s person but allows for 'a moderate amount of smoking';[34] in both cases, if smoking increases, more fresh air will be required. The figures cited should be 'adequate' to deal with odour control and to provide a reasonable indoor air quality. The pollutants both outside and inside obviously vary, and individual responses are subjective.

The issue of fresh air requirements has been examined carefully in the context of sick building syndrome (SBS). Sick building syndrome was first thought to be primarily a design problem affected by physical features such as the type of ventilation system or the depth of a space; it is now thought that management and maintenance are important.[35] One factor in increasing occupants' satisfaction is to include an ability on the part of the building, i.e. both the human systems and the physical constructions, to respond quickly to requests for change from its users.[36] An aspect of this is a degree of user control as provided by openable windows, adjustable blinds and manually adjustable thermostats.

Occupants make subtle choices between high temperatures, poor air quality and excess noise which are difficult for mechanical systems to match.[37]

Research into health and sick building syndrome has shown that indoor surface pollution – consisting not only of dust but of skin scales, debris from shoes and clothing, the products of smoking, eating and drinking and a number of other sources – seems to be one of the causes of the syndrome.[38]

House mites feed on skin scales and hair. Mites and their faeces are small enough to be suspended in air that is inhaled and can then affect people adversely. Fabrics such as carpets and upholstery are difficult to clean fully and serve as refuges for house mites. The problem of mites and dust is exacerbated by vacuum cleaning. If the cleaner's filter system is not fine enough to remove them, the smallest particles are lifted from the carpet where they are harmless, and are circulated into the room air by the jet of warm air rising from the cleaner's discharge. Solutions offered to this are improving filtration on cleaners, extracting all cleaners' discharges to outside, or, more drastically, removing soft furnishings.

Research continues – a recent study of buildings where concern had been expressed about air quality analysed people's complaints about odours and effects on health.[39] Everyday sources emitting volatile organic compounds were often involved; materials included white spirits from injected damp-proof courses, naphthalene from damp-proof membranes and 2-ethylhexan-1-ol from floor-coverings.

2.8 Moisture

Water and water vapour can cause problems in buildings for two main reasons. Firstly, a number of organisms, whose effects can be detrimental or damaging, can live in buildings and their growth is favoured in humid conditions. Mould, for example, will grow on surfaces that are consistently above 70% r.h.[40]

Secondly, water and high humidity levels can damage the materials used in buildings; for example, exposure of the steel in reinforced concrete to moisture could lead to corrosion and eventual failure.

Sources of water in buildings include rain penetration, rising damp, leaking pipes and the condensation of water vapour in the air. Breathing, washing, cooking and drying all release water vapour.

There are two common mechanisms that remove water vapour from the air in a building:

Dilution

If the moisture content of the outside air is less than that inside, ventilating with outside air will dilute and reduce the internal water vapour concentration. Cold air cannot carry as much water vapour as hot air, so it is generally drier than the internal air, even if it is 100% saturated. (In hot, humid weather the external moisture content may be higher than that inside so external air may in fact increase the internal moisture content.)

Condensation

If air is cooled, the water-carrying capacity of the air is reduced. At some point the water vapour in the air will reach 100% saturation (the dewpoint) and condense into water (Appendix A). Outside, this commonly occurs at night when the ground loses heat to the night sky and the water vapour in the air condenses as dew. (Air-conditioning systems employ the principle of condensation for dehumidification by passing air over cold metal coils.)

Inside, condensation will form on surfaces that are cooler than the dewpoint of the surrounding air. This can occur in a number of different circumstances. In low-income housing with high occupancies producing a great deal of water vapour and limited money for heating bills, condensation may occur on cold, poorly insulated walls.

In a well-ventilated, well-heated, lightly occupied building such as a stately home, one is more likely to get complaints of the air being too dry and the antiques cracking as a result. For example, see the discussion of Sutton House in Chapter 15.

In other buildings the surface of a chilled air duct or a cold water pipe can cause condensation, which can be avoided with a combination of measures:

- reducing the amount of moisture being released within the building;
- removing moisture from the air;

- ensuring that the surfaces are warmer than the dewpoint by a combination of insulating the walls and heating the space.

Preventing condensation at all times can be difficult but may be unnecessary providing that the space has a chance to dry over a daily cycle.[41] For instance, mould will not grow in a bathroom if the condensation due to bathing clears and the moisture content is reduced to below 70% r.h. after use.

Condensation within an element of construction is known as interstitial condensation. It is related to the temperature gradient in the element, say, a wall, which separates a space at one temperature and the outside at another. If water vapour flows through the wall (driven by the difference between the vapour pressures inside and outside) and reaches a part of the construction that is below the dewpoint temperature, condensation will occur. To prevent this, vapour checks are often used in wall constructions to limit the amount of water vapour flowing and to control the vapour pressure distribution in the wall. Repeated and prolonged interstitial condensation can lead to structural problems and must be avoided. Another condensation problem sometimes encountered is due to inadequate air circulation in a room. For example, if a wardrobe is put against a poorly insulated external wall the air between the two will be stagnant and cool. This, in turn, can lead to high relative humidities and condensation, and localized mould growth.

Guidelines

1. Comfort is important for the human body. Designers can increase the likelihood of thermal comfort in their buildings and their acceptability to occupants by providing user control, allowing for ample ventilation and limiting summer-time peak internal temperatures.
2. Both external and internal noise need careful consideration when developing designs.
3. Energy efficient buildings help to improve external air quality and to reduce the effects of global warming.
4. Control of relative humidity and condensation in buildings is essential in building design.

References

1. Humphreys, M.A. and Nicol, J.F. (1971) Theoretical and practical aspects of thermal comfort. Current Paper 14/71. BRE, Garston.
2. Hughes, J. and Beggs, C. (1986) The dark side of sunlight. *New Scientist*, **111**(1522), 31–5.
3. Pratt, A.W. (1958) Condensation in sheeted roofs. National Building Studies Research Paper No. 23. HMSO, London.
4. Anon. (1988) *CIBSE Guide, Volume A: Thermal Properties of Building Structures*, CIBSE, London.
5. Everett, A. (1975) *Materials*, Batsford, London.
6. Anon. (1993) *ASHRAE Handbook – Fundamentals*, ASHRAE, Atlanta, p 3.8.

7. Anon. (1992) Thermal environmental conditions for human occupancy: ASHRAE Standard 55-1992. ASHRAE, Atlanta.
8. Bunn, R. (1993) Fanger: face to face. *Building Services*, **15**(6), 25–7.
9. See reference 7.
10. Anon. (1966) Window design and solar heat gain. BRS Digest 68 (second series). BRS, Garston.
11. Anon. (1981) *Guidelines for Environmental Design and Fuel Conservation in Educational Buildings*, Department of Education and Science, London.
12. Petherbridge, P., Milbank, N.O. and Harrington-Lynn, J. (1988) *Environmental Design Manual*, BRE, Garston.
13. Taylor, A. (1987) Curing window pains. *Energy in Buildings*, **6**(6), 21–4.
14. Anon. (1987) *ASHRAE Guide: HVAC Systems and Applications*, ASHRAE, Atlanta.
15. Moon, P. (1940) Proposed standard solar radiation curve for engineering use. *Journal of Franklin Institute*, November, p. 604.
16. Lashof, D.A. and Ahuja, D. (1990) Relative contributions of greenhouse gas emissions to global warming. *Nature*, **344**(6266), 529–31.
17. Collingbourne, R.H. (1966) General principles of radiation meteorology, in *Light as an Ecological Factor* (eds R. Bainbridge, G.C. Evans and O. Rackham), Blackwell, Oxford.
18. Anon. (1989) Statutory Instruments 1989 No. 1790. Health and Safety, The Noise at Work Regulations 1989. HMSO, London.
19. Anon. (1986) EEC Council directive 86/188 on the protection of workers from the risks related to exposure to noise at work. *Official Journal of the European Communities*, 12 May 1986. HMSO, UK.
20. Anon. (1988) *CIBSE Guide A1: Environmental Criteria for Design*, CIBSE, London.
21. Anon. (1975) *Acoustics in Educational Buildings*, Department of Education and Science, Bulletin 51. HMSO, London.
22. See reference 11, p. 5.
23. Anon. (1988) Insulation against external noise. BRE Digest 338. BRE, Garston.
24. Anon. (1993) *ASHRAE Handbook – Fundamentals*, Chapter 11: Air Contaminants. ASHRAE, Atlanta.
25. Anon. (1986) *CIBSE Guide B3: Ventilation and Air Conditioning (Systems and Equipment)*, CIBSE, London.
26. Ibid.
27. Ibid., pp. B3–24.
28. Read, R. and Read, C. (1991) Breathing can be hazardous to your health. *New Scientist*, **129**(1757), 34–7.
29. Bown, W. (1994) Dying from too much dust. *New Scientist*, **141**(1916), 12–13.
30. Gorham, E. (1994) Neutralizing acid rain. *Nature*, **367**(6461), 321.
31. Mayo, A.M. and Nolan, J.P. (1964) Bioengineering and Bioinstrumentation, in *Bioastronautics* (ed. K.E. Schaeffer), Macmillan, New York.
32. Anon. (1991) Code of practice for ventilation principles and designing for natural ventilation. BS 5925 : 1991. British Standards Institution, London.
33. See reference 20, pp. A1–9.
34. Anon. (1989) Ventilation for acceptable air quality. ASHRAE Standard 62-1989. ASHRAE, Atlanta.
35. Leaman, A. (1994) Complexity and manageability: pointers from a decade of research on building occupants. *Proceedings of the National Conference of the Facility Management Association of Australia, Sydney*.
36. Ibid.
37. Anon. (1993) What causes discomfort? *Building Services*, **15**(6), 47–8.
38. Raw, G.J. (1994) The importance of indoor surface pollution in sick building syndrome. BRE Information Paper 3/94. BRE, Garston.
39. Anon. (1994) BRE tests the air. *BRE News of Construction Research*. April, p. 2.
40. Anon. (1985) Surface condensation and mould growth in traditionally built buildings. BRE Digest 297. BRE, Garston.
41. Ibid.

Further reading

Addleson, L. and Rice, C. (1991) *Performance of Materials in Buildings*, Butterworth-Heinemann, Oxford.

Anon. (1988) Sound insulation: basic principles. BRE Digest 337. BRE, Garston.

Anon. (1993) *ASHRAE Handbook – Fundamentals*. Chapter 8: Physiological principles and thermal comfort. ASHRAE, Atlanta.

Anon. (1993) *ASHRAE Handbook – Fundamentals*. Chapter 11: Air contaminants. ASHRAE, Atlanta.

Bartlett, P.B. and Prior, J.J. (1971) The environmental impact of buildings. BRE Information Paper 18/91. BRE, Garston.

Fry, A. (ed.) (1988) *Noise Control in Building Services*, Pergamon, Oxford.

Getz, P. (ed.) (1986) *The New Encyclopedia Britannica*, Encyclopedia Britannica, Chicago. Selected articles on convection, heat transfer, etc.

Gribbin, J. (1988) The ozone layer. *New Scientist*, **118**(1611), 1–4.

Gribbin, J. (1988) The greenhouse effect. *New Scientist*, **120**(1635), 1–4.

Gribbin, J. (1989) Quantum rules, OK!. *New Scientist*, **123**(1682), 1–4.

Buildings and energy balances

CHAPTER 3

3.1 Introduction

In this chapter the importance of buildings to energy use and carbon dioxide production in the UK is indicated. Energy use in different types of buildings is also examined in preparation for reducing it through design in subsequent chapters.

3.2 Buildings in the broad context

Buildings use energy and, as most of the UK's energy comes from the combustion of fossil fuels, produce CO_2 in the process. Delivered energy is the energy in fuels at their point of use (Chapter 7). Figure 3.1 shows delivered energy use and carbon dioxide emissions in the UK. Buildings account for about 45–50% of delivered energy use and just under 50% of all CO_2 emissions. The UK itself is responsible for about 3% of global CO_2 emissions.

Approximately 60% of building-related CO_2 emissions is due to the domestic sector and about 30% is attributable to the service sector, i.e. UK public and commercial buildings.[2] In the service sector the total CO_2 emission is about 89

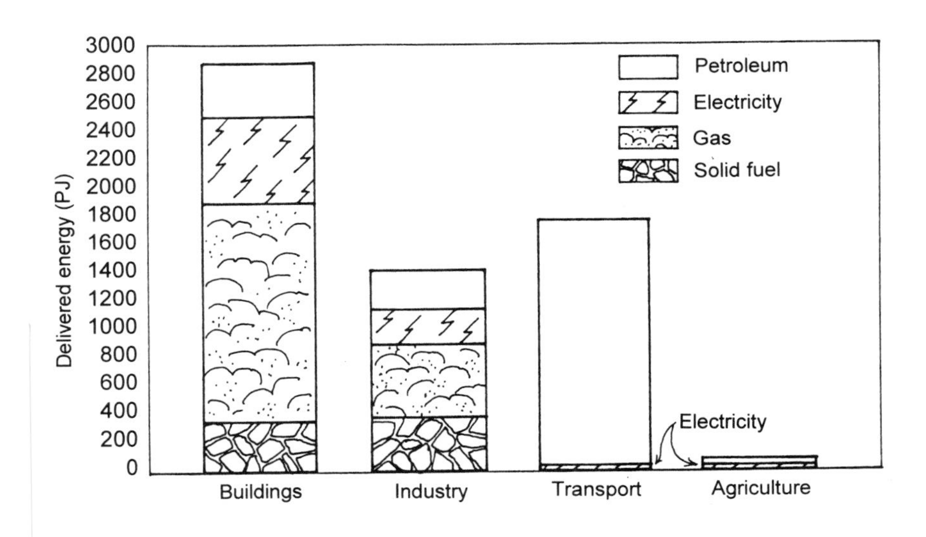

3.1(a) United Kingdom delivered energy consumption by sector and by delivered fuel type (1987).[1]

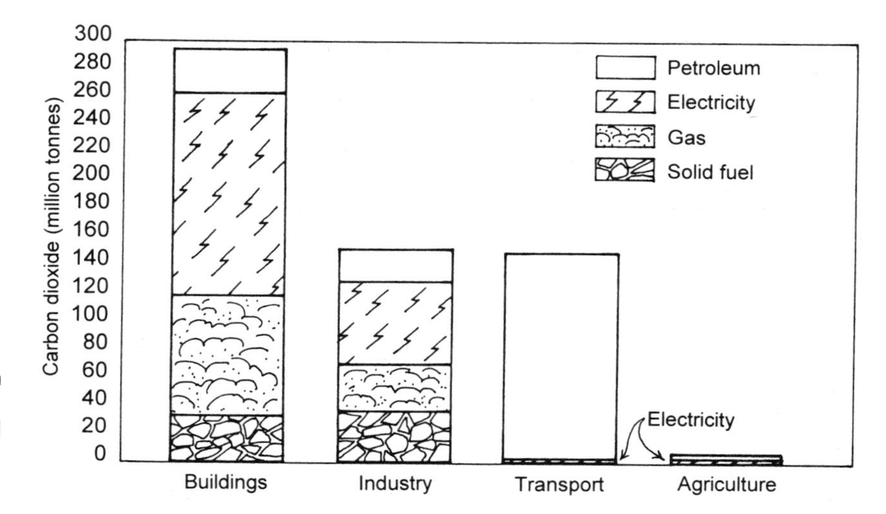

3.1(b) United Kingdom carbon dioxide emission by sector and by delivered fuel type (1987).[1]

million tonnes and approximately 44% of this is due to space heating (Table 3.1). Architects can have a major input through design on reducing these figures.

While architects can influence new building design fairly easily, it must be recognized that the vast majority of the building stock is existing. To reduce energy consumption the insulation of these buildings needs to be upgraded, but in practice this can be difficult. Increasing the loft insulation of housing has proved to be one of the easier approaches so far.

Table 3.1 UK carbon dioxide emissions by end use for the UK service sector[3]

Use of fuel	%
Space heating	44
Water heating	7
Lighting	17
Cooking	6
Air conditioning	6
Refrigeration	7
Power	13

3.3 Energy flows in buildings

In Chapter 2 we saw that energy exchanges affect people and buildings. Externally, the building envelope is subject to solar gain, radiation exchange with its surroundings and convective heat loss (or gain) owing to the winds that almost

Table 3.2 Approximate energy consumption and carbon dioxide production for selected activities, equipment and buildings[a]

Item	*Energy consumption (kWh)*	*CO_2 production (kg)*	*Period*	*Notes*
Man at rest	2.8		day	
Shower	1.8	0.4	5 minutes	Water heated with gas
Bath	3.3	0.7		80 litre bath heated with gas
Dishwasher	2	1.5	1 cycle	Including heating the water with electricity
Fridge/freezer	2.2	1.6	24 hours	
100-watt filament light bulb	2.4	1.8	24 hours	
Equivalent miniature fluorescent	0.5	0.4	24 hours	
For an average 3-bedroom dwelling:				
Electricity consumption	2 000	1 500	1 year	Excluding water or space heating
Water heating	2 000	420	1 year	Using gas
Space heating	22 500	4 720	1 year	Using gas
Space heating – new house target	10 000	2 100	1 year	Using gas
Domestic heat recovery ventilation system	1 750	1 300	1 year	Electricity in running the fans
For a 1500 m² primary school:				
Electricity use	45 000	34 000	1 year	Mostly lighting
Gas use	225 000	47 000	1 year	Mostly heating
For a 1500 m² air-conditioned office				
Electricity use	525 000	395 000	1 year	
Gas use	420 000	88 200	1 year	
100 km car journey	100	22	2 hours	
1 acre of corn		3 000	1 year	CO_2 fixed

[a]Figures are indicative and can often be improved with design.

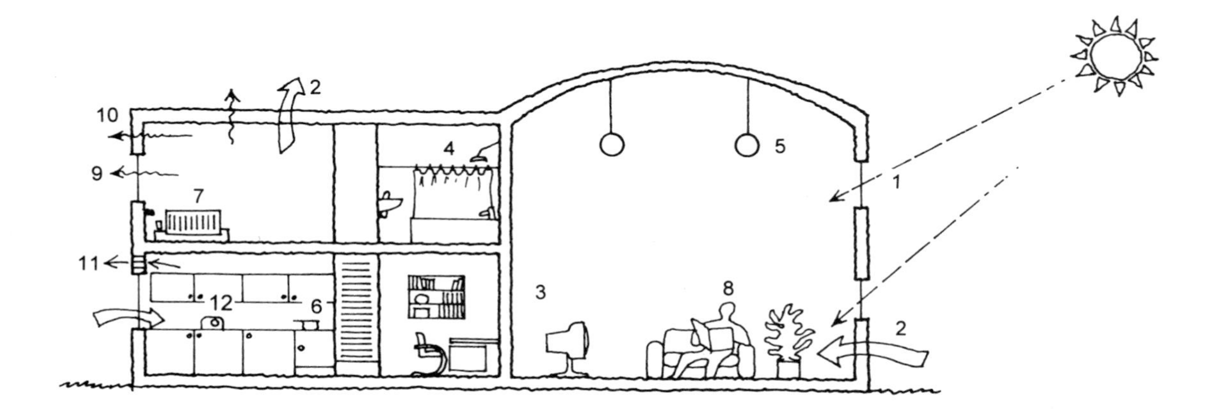

Key
1. Solar gain 200–600 W/m^2
2. Fresh air 10 l/s 0 °C to 21 °C = 250 W
3. TV 100–400 W
4. Shower 1.5 l/s, ΔT = 30 °C = 18 900 W
5. Lighting: domestic lamps are 60–100 W, small fluorescents are 11–17 W
6. Cooker rings 500–1500 W. Most of the heat goes into turning water into steam. Oven 1000–3000 W
7. Radiator 1000 mm long x 490 mm high (single panel): 670 W
8. People 80–300 W (higher value when exercising)
9. 1m^2 window heat loss
   110 W/m^2 single glazing
   50 W/m^2 double glazing
   40 W/m^2 double glazing low-emissivity coating
   Assuming a 20 K temperature difference between inside and out
10. Building fabric
   40 W/m^2 if solid brick
   10 W/m^2 wall with 100 mm of insulation
   5 W/m^2 wall with 150 mm of insulation
   Assuming a 20 K temperature difference between inside and out
11. Extract fan 60–120 W
12. Radio 4 W

3.2 Approximate energy flows in a house (Sunday at home).

continuously flow past it. Moisture, too, can be lost to the surroundings (and, somewhat differently from the human body, gained as when driving rain penetrates into the external wall). Internally, the building is the site of the activities of the occupants and processes which include lighting, running of equipment from ventilation fans to photocopiers, cooking and heating. All of the energy supplied finishes as heat, even when there is a useful intermediate stage such as light or a moving fan.

Table 3.2 shows some typical energy consumption figures for a variety of processes and the associated CO_2 production.

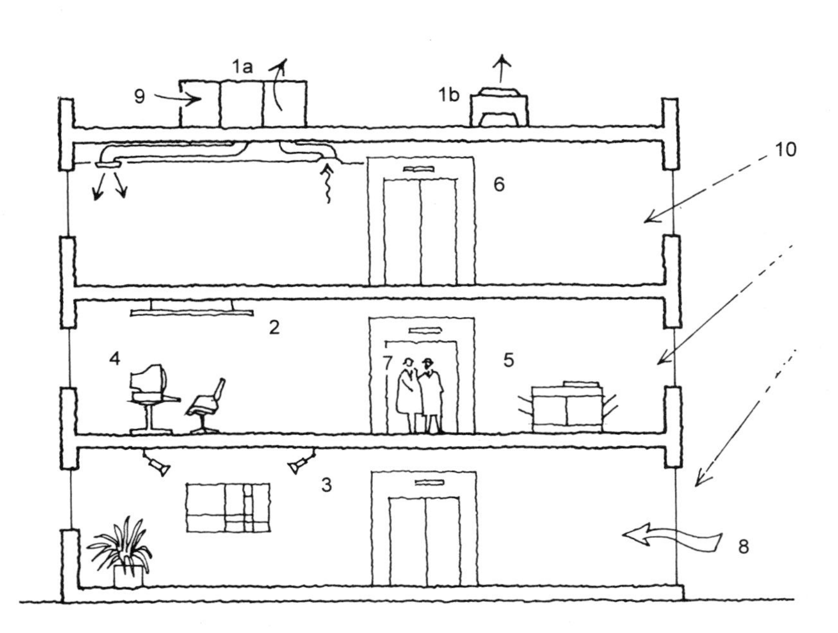

Key

1.	Air conditioning (a) air-handling unit (b) heat rejection equipment	10–30 W/m² of office for total plant electrical load including chiller (not shown)
2.	Fluorescent lighting	10–20 W/m² of office to give 300–500 lux
3.	Tungsten lighting	20–100 W/m² of office to give 100–300 lux
4.	Computer	50–250 W
5.	Photocopier	1000–5000 W
6.	Lift	10 000 – 30 000 W
7.	Person	100–140 W
8.	Warm air entering in summer through open door	2000 W (1 m³/s)
9.	Fresh air	150 W per person at 10 l/s person
10.	Solar gain	200–600 W/m² of window

3.3 Approximate energy flows in an air-conditioned office with summer cooling loads.

Figures 3.2 and 3.3 show typical energy flows in a house and an air-conditioned office.

If we first consider the domestic situation, we see in Figure 3.2 that there is an input from the Sun which contributes to heating and lighting and which, although free, is highly valuable.

The advantages of solar gains need to be balanced against potential overheating and heat loss through glass. This issue also obviously applies to offices and other building types. It illustrates the need to view the building as a system and at times to balance contradictory factors. If one element is given too much emphasis – for example, daylighting in system-built schools in the UK in the 1960s – problems can result. In the summer such schools suffered from overheating and

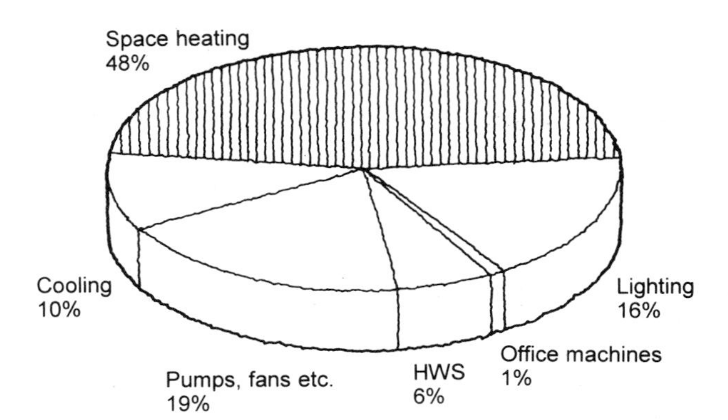

3.4 Delivered energy use in a typical air-conditioned office building.[5]

glare, and in the winter they had high heat losses. If, on the other hand, form, glazing, fabric, ventilation and services work together, comfort and energy efficiency can be achieved.

Many loads in buildings are highly intermittent, such as hot water use for showers or washing hands, or energy use for cooking. Others, such as heating and lighting, can be more constant.

Loads need to be examined and quantified so as to be able to identify how best to provide the energy required and how to reduce the demand as much as possible. Lighting levels in a domestic living room are likely to be much lower than in offices, but because the office lights tend to be much more efficient the actual electricity consumption may be similar.

In a domestic situation about 23% of the total energy requirement is for heating water (for handwashing, bathing, cooking and so forth).[4] Much of the heat is lost down the drains or goes into evaporating water which is added to the internal air.

Ventilation provides oxygen, fresh air, removes CO_2 and odours and helps prevent condensation problems by taking away water vapour. Heating systems replace heat loss through the fabric (which is falling as insulation standards increase) and heat the ventilation air in the winter.

The heat requirement for the ventilation air supply becomes more and more important as fabric heat losses fall. (Chapter 9 discusses the possibility of using mechanical ventilation systems to supply and extract ventilation air and to recover some of the heat in the extract air.)

Most air-conditioned buildings are completely sealed off from the outside and air is supplied from a central air-handling plant (shown schematically in Figure 3.3). Heat from the Sun, equipment and people must be removed by air from the central plant and this necessitates a cool air supply and so requires energy. Figure 3.4 gives typical data.

Energy for cooling (and associated equipment) and lighting are two areas where significant reductions can be made, and we shall examine some approaches in later chapters.

Guidelines

1. Buildings are major consumers of energy and producers of CO_2. Architects and engineers can help change this.
2. Analysis of energy use and flows helps identify where savings can be made.

References

1. Henderson, G. and Shorrock, L.D. (1990) Greenhouse-gas emissions and buildings in the United Kingdom. BRE Information Paper 2/90. BRE, Garston.
2. Shorrock, L. (1994) Private communication.
3. Moss, S.A. (1994) Energy consumption in public and commercial buildings. BRE Information Paper 16/94. BRE, Garston.
4. Taylor, L. (1993) *Energy Efficient Homes: A Guide for Housing Professionals*, Association for the Conservation of Energy and the Institute of Housing, London.
5. Anon. (1991) *Energy Audits and Surveys: CIBSE Applications Manual AM5*, CIBSE, London.

CHAPTER 4

Building planning and design

4.1 Introduction

This chapter covers shape and size, the 'body' and the 'skin' of the building and issues of internal organization. It provides a basis for articulating the building on the site in order to provide an energy efficient and comfortable internal environment.

4.2 Form

The orientation of a building may be fixed but if choice is possible it should face south to take advantage of the Sun's energy (Chapter 5). Total volume, too, is likely to be prescribed and so, often, the first major design decisions are allocating volumes to various activities and developing the form of a building.

Form is governed by a number of functional considerations that are discussed below, and in more detail in the following chapters, and include:

- the use of the Sun's energy and daylight (Chapters 5 and 8)
- provision of views for occupants
- heat loss through the building envelope
- the need for ventilation (Chapter 9)
- acoustic attenuation if required.

In the recent past, the glass blocks of Mies Van der Rohe epitomized an architecture that shut out the natural environment and provided an acceptable internal environment through the use of considerable energy and sophisticated services.

The Queens Building at De Montfort (Chapter 13) is the antithesis of this and articulates the building both on plan and in section to respond to the environment and make the best use of natural energy sources. The likelihood is that environmental considerations will allow for freer forms and, thus, a welcome architectural diversity; but before we can draw any conclusions about form we need to know more about how buildings work.

4.3 The building 'body'

An important consideration is how quickly a building responds to heat inputs (internal and external), and this is related to the thermal conductivity of its materials, the thermal mass (or heat capacity – both discussed above in Chapter 2), and, related to these, the admittances of the elements of the construction.

The admittance, *Y*, of a constructional element, put simply, is the amount of energy entering the surface of the element for each degree of temperature change just outside the surface and, as such, has the same units as the U-value (W/m^2 K) (Appendix B). The admittance of a material depends on its thickness, conductivity, density, specific heat and the frequency at which heat is put into it. (In addition to the admittance, the response of building elements to energy cycles depends on the decrement factor and the surface factor;[1] put simply, once again these factors are associated with time lags in energy flows, with the decrement factor representing the 'damping' effect of an element's response to an energy gain.) Considerably more technical explanations of these concepts are to be found in References 1, 2 and 3. Table 4.1 gives properties of some constructions.

Table 4.1 Admittance and density of selected construction elements[4]

Item	*Admittance (W/m^2 K)*	*Density (kg/m^3)*
1. 220 mm solid brickwork, unplastered	4.6	1700
2. 335 mm solid brickwork, unplastered	4.7	1700
3. 220 mm solid brickwork with 16 mm lightweight plaster	3.4	1700 for brickwork 600 for plaster
4. 200 mm solid cast concrete	5.4	2100
5. 75 mm lightweight concrete block with 15 mm dense plaster on both sides	1.2	600 for concrete 600 for plaster

As can be seen from the table, dense constructions have higher admittances, which is to say they absorb more energy for a given change in temperature. (One must be careful, however, because for multilayer slabs, the admittance is determined primarily by the surface layer; thus, a 300 mm slab with 25 mm of surface insulation will respond more as a lightweight than as a heavyweight material).[5]

If a building absorbs a great deal of heat and only experiences a small temperature rise it is said, in no very precise manner, to be thermally heavyweight. Such buildings tend to have high admittances and a great deal of thermal mass, usually in the form of exposed masonry. Lightweight buildings, on the other hand, may have thin-skinned walls, false ceilings with lightweight panels, metal partitions and so forth. The CIBSE[6] has tried to be more precise and has defined a heavyweight building as one whose ratio of admittance value to U-value is greater than 6; British Standard 8207,[7] on the other hand, uses a ratio of 10. The concept matters more than the number.

The particular importance of these issues is in providing comfortable conditions in the summer without the use of air conditioning. This is not simply a problem for office buildings – countless schoolchildren in the UK were educated in the 1960s and 1970s in lightweight, underinsulated, overglazed buildings that overheated in the summer, particularly on the top floor in westerly-facing classrooms in the late afternoon.

Normally, the heat flow into a building from the outside is approximately cyclical. On a daily basis, the Sun rises, the air temperature increases and heat is transferred directly via windows and indirectly via the building structure. As the Sun sets the building starts to cool, and the following day the cycle continues. In the winter, the external gains are insufficient and so the heating system supplies heat each day during the period of occupancy. At night, the temperature is allowed to drop to conserve energy. Again, the following day the cycle continues.

The thermal mass of the building evens out the variations. In the summer, by delaying the transfer of heat into a building, the time the peak temperature is reached can be altered. By using high-admittance elements the building fabric can store more of the heat that reaches the internal and external surfaces, thus reducing the peak temperatures. This 'balancing' effect can apply both during the day and at night, because if cool night air is brought into contact with high-admittance surfaces their temperatures will drop, i.e. there will be cool thermal storage. The next day, when warmer day-time air flows over the same elements, they will be cooled thus improving comfort conditions for the occupants. This technique is used both at RMC (Chapter 11) and De Montfort (Chapter 13).

Architecturally, the key requirement is to incorporate high-admittance materials in the building and expose them in an appropriate manner. This means that false ceilings, raised floors and plastered walls will need to be kept to a minimum. Appearance will obviously be an important consideration as walls and soffits (and services) are bared. However, there are a number of solutions – from brickwork walls with coloured bands to make them more attractive, to high-admittance ceiling linings such as cement-bonded chipboard. It may also be possible to exploit more complex approaches such as taking the incoming air supply over a concrete floor slab. Greater floor-to-ceiling heights will also, of course, provide more thermal mass for a given floor area. In some cases one element may be made to perform several functions. At De Montfort the heavy masonry stacks ventilate, provide thermal mass and help support the roof.

Heavyweight buildings have an important role to play where air-conditioning might otherwise be needed. However, study of a number of buildings has shown that:

- if loads are low, there is a limit to the need for thermal mass, and
- there can be a limit to its usefulness.[8,9]

To make efficient use of mass, one must be able to ventilate at night to lower the temperature, otherwise the heat absorbed tends to accumulate and discomfort results. This has practical implications: if night-time ventilation is under automatic control, the system should not be too complex; if under manual control, it needs to be foolproof, both in maintaining security and in preventing the entry of rain.

If loadings are low and air movement is good, comfort can be achieved with lightweight buildings. If a building is always in use – for example, sheltered housing schemes – heavyweight buildings are often appropriate, but if occupancy is intermittent a lightweight building can have a positive advantage. For example, in winter, the heat stored in a heavyweight building during the day may be released at night when there is no need for it. The process is somewhat similar to an electric storage heater that supplies heat during the day when needed, but cannot stop releasing heat after people have left. The significance of this, however, depends on the building; and as insulation standards increase and buildings become better sealed, there is a decrease in the amount of heat wasted by a building when all the occupants have left.

Unfortunately, there are no definite rules; each building needs to be examined on its own merits, and we shall return briefly to these considerations later in this chapter.

4.4 The building 'skin'

Development of the building envelope, or 'skin', is likely to be rapid in the next decade or so. Technological innovation in glass will allow window systems to respond to environmental conditions in ways not previously commercially viable for buildings. Sun-glasses which react to different light conditions are but a hint of the potential of glass.

Building envelopes obviously need to be durable, economical, aesthetically pleasing, weathertight, structurally sound and secure. Psychologically, views out are very important. Environmentally, the questions that need to be addressed are: how they respond to solar radiation (both for the Sun's heat and light), how ventilation is made possible, how heat loss is minimized and how noise is controlled. The envelope will, to a large extent, determine how the internal environment is affected by the external one.

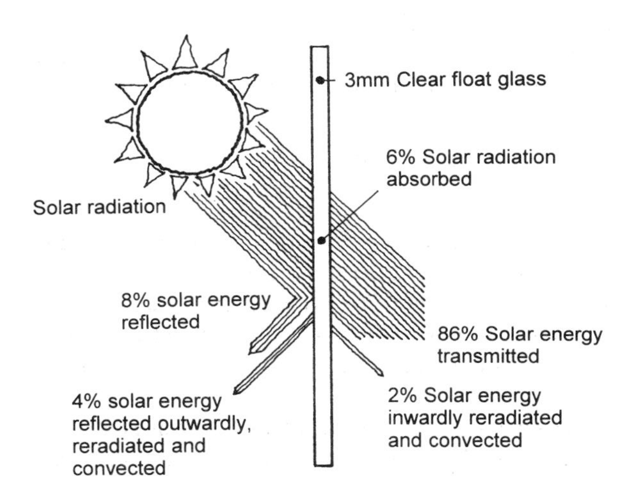

4.1 Energy exchange at a window of 3 mm float glass.[10]

Solar radiation

Figure 2.4 shows the spectral distribution of solar radiation, to which the components of the envelope react in different ways. If we first consider the opaque elements, the amount of radiation absorbed at the surface depends in part on the colour of the surface. Lighter colours, of course, absorb less and reflect more of the incident radiation (Table 2.2).

Turning to translucent materials each one has a different characteristic. Figure 4.1 shows the energy exchange for plain 3 mm float glass. The percentage of solar radiation transmitted by a window varies with wavelength, as shown in Figure 4.2.

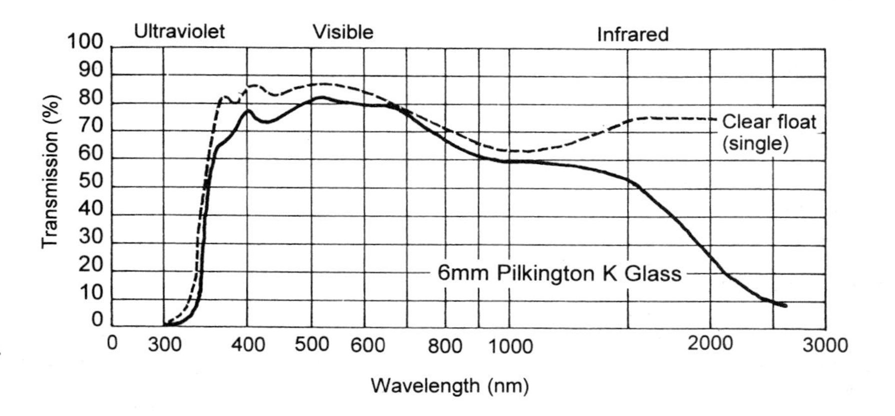

4.2 Spectral transmission curves for glass.[11]

Figure 4.2 shows that glass filters the Sun's radiation much as the atmosphere does, absorbing some of the UV and infrared and letting through much of the visible light. A glasshouse will let in a great deal of solar radiation but will not transmit much of the far infrared produced by the room, much as clouds block the Earth's outgoing radiation. (See Figure 2.3 for an approximate spectrum of room radiation. Figure 4.2 does not continue far enough to the right to show the reduced transmission of clear float glass at longer wavelengths.)

The amount of radiation that enters and exits a room can be controlled to a certain extent by altering the components of the glass, by using several layers of glazing, by applying special coatings and filling the spaces between the panes with various gases, or by evacuating them; an example of the altered transmission characteristics is seen in the graph in Figure 4.2.

The heat loss from any building element is related to its U-value (Appendix B). U-values for different glazing types along with transmission and acoustic characteristics are shown in Table 4.2. (Note that this is for glazing alone; a more precise analysis would be needed to take the frame into account.)

The table shows that there is some loss of light and solar radiant heat as the U-value improves. However, in most applications this is not a significant disadvantage compared with the benefits obtained. It also shows that direct solar transmittance is not the same as direct light transmittance, and this suggests possibilities for glass development. In the summer, for example, an ideal glass would

Table 4.2 Characteristics of glazing systems[12]

Type	*U-value (W/m^2 K)*	*Light transmittance*	*Solar radiant heat transmittance*[a]		*Mean sound insulation*[b]
			Direct	Total	(dB)
Single (4 mm clear float glass)	5.4	0.89	0.82	0.86	28
Double glazing (6 mm clear float inner, 12 mm airspace,[c] 6 mm clear float outer)	2.8	0.76	0.61	0.72	30
Double with low emissivity coating (6 mm Pilkington K inner, 12 mm airspace, 6 mm clear float outer)	1.9	0.73	0.54	0.69	30
Double with low emissivity coating and cavity (6 mm Pilkington K inner, 12 mm airspace with argon, 6 mm clear float outer)[13]	1.6	0.73	0.54	0.69	30

[a]Direct solar radiant heat transmittance covers the entire solar spectrum of approximately 300–2200 mm. Total solar radiant heat transmission is the sum of the direct transmittance plus the proportion of absorbed radiation re-radiated inwards.
[b]Mean sound insulation is for the frequency range of 100–3150 Hz.
[c]At airspaces above 12 mm the U-value is about the same. Below 12 mm it gets worse and typical U-values for 6 mm and 3 mm gaps are 3.2 and 3.6 W/m^2 K, respectively.[14]

transmit light (to reduce the need for artificial lighting) but no other part of the solar spectrum (to keep the space cooler). In the winter both light and heat are likely to be advantageous. Similarly, in the winter a very low U-value saves energy. If, in the summer, the internal temperature is above the external – as often occurs in lightweight, non-air-conditioned buildings – a high U-value would help get rid of the heat. Glasses whose characteristics can be altered have enormous potential.

Energy loss through a window depends particularly on internal and external temperatures and is independent of orientation. Energy gain, on the other hand, obviously depends on direction because of the Sun. Appendix A gives a selection of solar data. When solar radiation data is used with internal and external temperatures it is possible to determine the daily solar heat gain and average conduction heat loss – the difference between the two is the daily energy balance. (Figure 4.3 shows energy balance data.)

Newer glasses are likely to have even more favourable energy balances but Figure 4.3 nonetheless presents a broad picture that energy is available and that we should be using it. In doing so we shall, of course, need to guard against overheating and glare. We shall return to this when we discuss shading devices.

First, however, it is worth while just mentioning some current areas of glazing

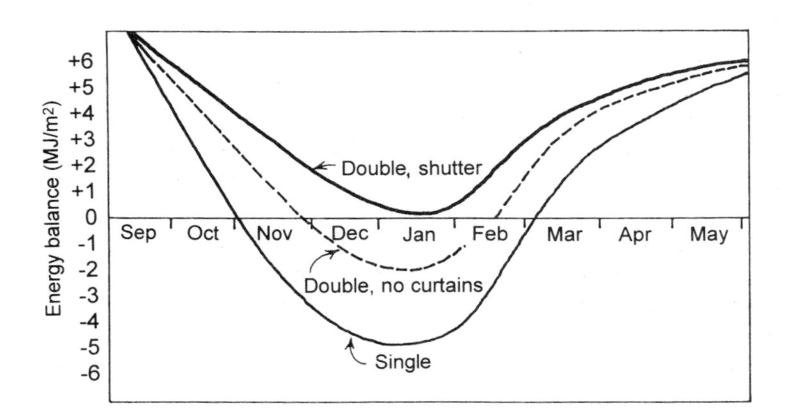

4.3 Daily energy balance for south-facing glazing at Bracknell, Berkshire; assumed internal temperature 18 °C.[15]

research. Thermochromic glasses include clear films which, when heated above a certain temperature, turn an opaque white and so can be used to reflect sunlight.[16] Electrochromic glass has layers whose properties change when a voltage is applied to them.[17] In this way parts of the solar spectrum can be reflected from the glazing thus reducing the heat that enters the building. Transparent (or more precisely, translucent) insulation materials include glass/aerogel/glass assemblies[18] and polycarbonate plastics which can be applied to the external walls of buildings.[19]

Shading is normally needed to control overheating in the summer and in some designs in the spring and autumn. Shading devices need to be considered as systems rather than isolated elements and shading control at the building envelope must be related to the activities in the building, its mass and its ventilation system. A historical example, illustrating the need for a systems view, is Le Corbusier's Salvation Army Hotel in Paris. The original design included a way of removing heat from in front of an inner skin of unopenable south-facing glazing. However, for cost reasons the design was altered leaving only the fixed glazing which almost roasted the occupants.[20] Later, a brise-soleil, or sun screen, was added to reduce overheating.

A disadvantage of external, fixed shading is that it results in some permanent loss of passive solar gain when needed. Note that this is true even if one attempts to design an arrangement based on direct solar gain which blocks out the Sun's

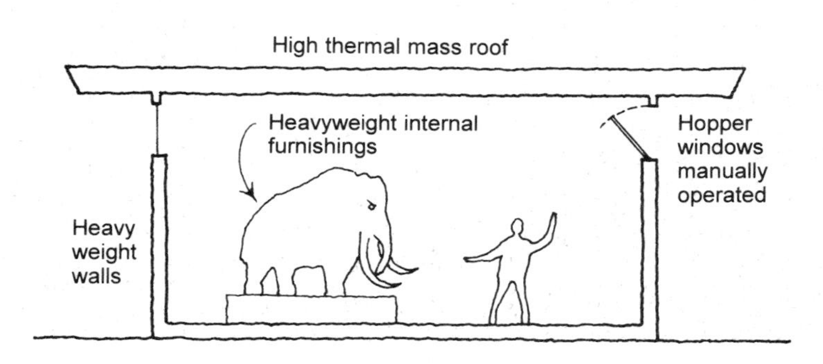

4.4 External shading and night-time ventilation.

rays in June but allows them entry in December. There is also a permanent loss of daylight with fixed external shading. Nonetheless, architects are often drawn to it because it can enliven an otherwise banal façade.

It is, nevertheless, too easy to say 'avoid external shading' and it is much better to examine the functional requirements. Structural overhangs are one form of shading that also offers the possibility of rain-shielding, as shown in Figure 4.4, which is based on the Regional Museum of Prehistory at Orgnac l'Aven in France.

In this situation the required light levels were low. A combination of overhang and hopper window means that very simple night ventilation can be provided without major risk from rain entering if someone forgets to close the windows; the night ventilation works well with the high thermal mass.

Movable external devices tend to be costly and because of exposure to the weather require significant maintenance. They should be used only after careful consideration.

Movable shading devices such as blinds placed between glazing layers let more heat into a space than external shades but are more reliable. They are also more effective than internal blinds (but more costly). Very approximately, if single glazing allows in 1.0 unit of solar energy, single glazing with internal blinds will allow 0.67 unit. Double glazing will allow 0.88 unit and double glazing with internal blinds, 0.33 unit.

Figure 4.5 shows 25 mm blinds between two panels of sliding glass at St John's College, Cambridge.

4.5 Venetian blinds in the cavity of a double window.

Internal shades include curtains, blinds (with wooden, metal or plastic slats) and shutters. Shutters may, of course, include slats. Curtains, blinds and shutters may incorporate thermal insulation with options varying from simple slabs of, say, mineral fibre built into shutters, to sophisticated aluminium and polyester layers in blinds.

Increasingly, internal shading devices also need to be considered in connection with ventilation (Chapter 9) and in some cases, such as lecture theatres and classrooms, the need to black out (or more precisely, grey out) a space.

Many of the considerations above apply equally well to daylight (Chapter 8). A résumé of shading devices is given in Reference 21.

The world is littered with inappropriate, dysfunctional, flimsy, unreachable, unmaintainable, unaesthetic shading devices. Beware!

Ventilation

Ventilation of buildings has varied from uncontrolled infiltration at cracks around windows, doors, junctions, floorboards and so forth for, say, homes, to purpose-made openings to provide air to air handling units in sealed air-conditioned offices. In most domestic situations it was assumed (usually correctly) that enough air would enter the rooms to meet the needs of oxygen, odour and pollutant removal, condensation control and, in the summer, possible removal of heat. Indeed, generally too much air gained entry during the winter and this led to excessive heat loss. Other problems also occurred but they tended to be localized, and were frequently the result of a combination of high moisture production, low temperatures and inadequate ventilation. The resultant condensation and attendant problems have been mentioned in Chapter 2.

For several years, partly as a result of increased interest in energy conservation, there has been a growing interest in ventilation, summarized by the slogan 'Build tight, ventilate right'. If the right amount of air is provided, if heating systems distribute heat as needed throughout the building, and if moisture is dealt with at the source, for example through extract fans in kitchens and bathrooms, there is a high probability that condensation will be controlled and energy consumption kept reasonably low.

But how to provide the 'right' amount of air? As a starting point, the building needs to be tightly sealed so that entry and exit points for air are controllable, or at least well defined. A tightly sealed construction requires careful design and good workmanship. Flexible sealants are required at junctions, say, between window frame and walls and at interfaces of steel frames and masonry; and when detailing external joints, allowance must be made for thermal expansion, deterioration, distortion and weathering. It is not uncommon now to pressure test buildings to ensure that they meet standards of air tightness; for example, a building might be required to have an hourly air change rate of no more than 0.1–0.2 at 50 Pa. Openings for ventilation will vary according to the application, and windows have normally been the main means of providing natural ventilation in both winter and summer. (Different types are discussed in Chapter 9.)

For small amounts of (permanent, not fully controllable) winter ventilation trickle ventilators incorporated into window frames have become popular, espe-

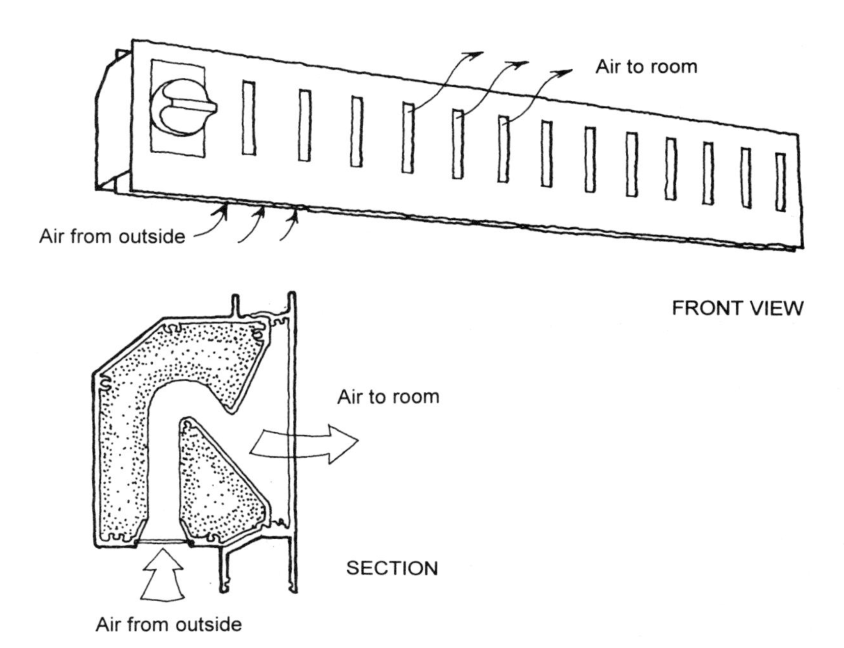

4.6 Trickle ventilators.

cially in domestic situations; the example shown in Figure 4.6 incorporates acoustic attenuation.

Ventilators are also available which incorporate temperature and humidity sensors and open at the set points, thus providing direct control.

Heat loss

Heat loss at the building envelope is principally a matter of the U-values of the glazed areas (Table 4.2) and the insulation used in the construction of opaque wall elements, roofs and (to a lesser extent) floor slabs. Insulants will be discussed in Chapter 6.

Noise

Control of noise is a key issue for many sites, and in many ways noise and acceptable air quality create more difficult problems than overheating due to solar gains. Progress in developing successful designs will have major implications for eliminating the need for air-conditioning and elaborate mechanical ventilation systems. Appendix C gives some of the basic terms and data needed to discuss the issues.

To give an idea of the scope of the problem in urban situations, it has been suggested that, for design purposes, an L_{10} value of 77 dBA be used for noise outside office and commercial buildings.[22] If we try to design for, say, an L_{10} value of 45 dBA inside we would need 32 dB of attenuation. Now let us consider the building envelope as being made up of the opaque solid elements, the glazing and the ventilation openings. Typical masonry walls (and many other kinds) have

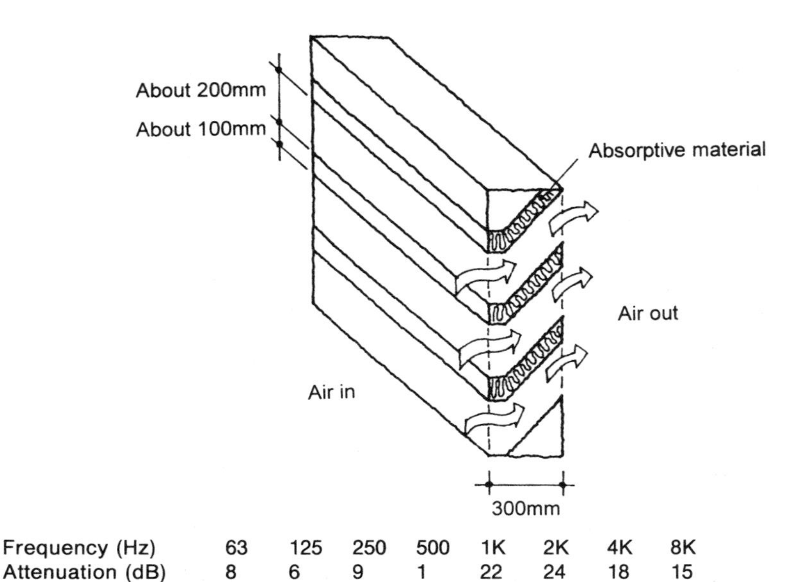

Frequency (Hz)	63	125	250	500	1K	2K	4K	8K
Attenuation (dB)	8	6	9	1	22	24	18	15

4.7 Cut-away view of acoustic attenuators and approximate attenuation performance.[24]

no difficulty (Figure 2.7) meeting this attenuation figure as long as they are well constructed, but double-glazing systems do not quite meet the requirement (Table 4.2). (To achieve very high values of attenuation it is necessary to increase the space between the panes significantly. A typical acoustic glazing unit might consist of 6 mm glass, a gap of 200 mm or more and a sheet of 10 mm glass with the panes of glass not parallel to each other and absorbent reveals; this could provide up to 45 dB attenuation.)

The weakest link, however, lies with openings for ventilation because, generally, where air can enter, noise can also enter. An open window has an average attenuation of about 10 dB.[23] An opening in a wall can incorporate acoustic attenuation in a variety of ways and the performance will vary accordingly. Figure 4.7 shows a typical solution and gives attenuation data. (Note that 'bending' the air path makes it more difficult for sound to enter directly.)

Performance is worst at low frequencies. This is generally true for attenuators and is one reason why most of us live with a background low-frequency rumble. The attenuation can usually be improved by incorporating bends in the air paths, but this increases the resistance to air flow and so makes it less suitable for natural ventilation systems with their low driving forces; however, some simple mechanical ventilation added to the natural ventilation can be of assistance (Chapter 9).

Of course, in many situations noise levels will not be very high and so attenuation will be less necessary. It may also be possible to draw air from a quieter area around the building (and which may also have a higher air quality). In other situations, some background noise will be desirable. In university study bedrooms, for example, eliminating external noise (whether from traffic, the wind or passers-by) can be disconcerting and will tend to exacerbate the irritation due to noise from adjacent study bedrooms. In open plan offices, as noted in Chapter 2, some background noise is usually desirable because it helps mask individual con-

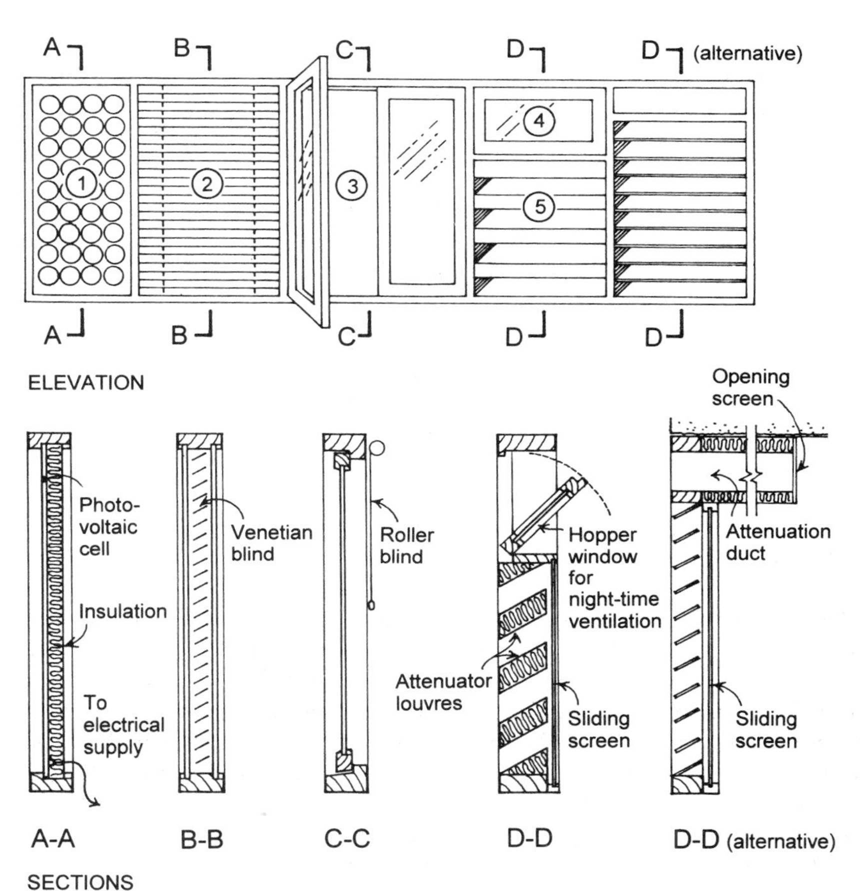

Key
1. Future power section, e.g. photovoltaic cells (Chapter 7).
2. Fixed glazing with mid-pane blinds: allows variable daylighting level and control of solar gain.
3. Openable glazing (numerous types available – casement window shown) allows ventilation (but admits noise).
4. Night-time ventilation from manually openable hopper window; provides security and protection from rain.
5. Opaque section provides ventilation and acoustic attenuation. Light penetration can be minimized. Additional attenuation can be provided at soffit level if required, as shown in section 'D-D alternative'.

4.8 Composite 'window'.

versations. Also, individuals often prefer to have some control over their environment and will choose a little more fresh air along with some more noise in preference to less noise but higher temperatures.

Each situation requires individual analysis. It does appear, however, that there is an argument for separating out the traditional functions of the window so that light, ventilation and noise can be dealt with more effectively. Too many 'solutions' are currently trying to do too much in too tight a space. Figure 4.8 shows a composite theoretical 'window' or, more accurately, wall which illustrates some options and future directions.

Obviously, possibilities and permutations are numerous. Where blackout facilities are required all blinds can be specified as such. If external solar control is desired it can easily be incorporated, and if light shelves (Chapter 8) or other such devices are needed they too can be added. As active noise controllers (systems which create equal and opposite pressure variations at the same frequencies) improve and become cost effective they may start to replace (or complement) the passive devices shown in the figure.

For more sophisticated (and costly) applications the functions shown can be motorized. Instead of a sliding screen, as shown in Section D–D, automatic dampers could be used and all opening windows could be motorized. These could be further linked to noise detectors or even odour detectors as these become commercially available in the future.

Window design, which is exceptionally important, is difficult and is treated in more detail in Chapter 9.

Having examined the main issues we can ask: What percentage of a wall should be glazed? Unfortunately, the answer is not straightforward. It depends on the building type (office or domestic), the building 'body', internal loads and so forth. In order to increase natural lighting in multi-storey offices, and thus reduce the energy consumption of artificial lighting, a large percentage of the wall needs to be glazed (Chapter 8). There are then at least two possibilities. The first is that adequate control at the building envelope plus natural ventilation pathways (with possible mechanical assistance) result in sufficient comfort. Alternatively, the space begins to overheat and the energy requirement for mechanical ventilation and, perhaps, air-conditioning, increases. Obviously, the first possibility is preferable – the principle to follow is designing to make the best use of available energy and incorporating controls to ensure that the advantages do not become nightmares. Unfortunately, there are no simple answers.

In domestic situations, where overheating is usually less likely, the annual space heating requirement has been studied as a function of the U-value of different glazing types and the percentage of south-facing glazing area to floor area for a 140 m^2 house.[25] If the glazing is not insulated at night, with a U-value of 2.0 W/m^2 K the space heating is minimized with very roughly 50% of the south wall glazed. If insulation is added the percentage can be higher and the space heating requirement falls.

Second 'skins', buffer spaces and atria

It is also possible, of course, to put a second 'skin' or envelope around the first. This can help reduce heat loss while maintaining most of the benefits of solar gain directly into the building. It can also trap potentially useful solar heat between the skins. Acoustic attenuation is a further potential advantage. Buffer spaces like conservatories are localized second 'skins'.

Disadvantages of a second envelope include cost, possible interference with natural ventilation and creation of a large enclosed void connecting the building's floors through which smoke and fire can spread. Any proposed design must carefully analyse these issues.

Other less complicated buffer zones include draft lobbies and auxiliary spaces such as garages. In these the environment is uncontrollable or only loosely controlled but protection is afforded to the primary space. Atria are 'buffer' spaces that can vary widely in complexity. In their simplest form they are enclosed spaces that keep the rain out, allow light and solar radiation to enter and have no artificial heating; these simplest forms are also likely to have high level openings or extract systems to dispel heat and smoke. Complex atria tend to have installed heating systems and ventilation systems that interact with the spaces around the atria. They can also incorporate solar and light controls.

Atria have the potential environmental advantages of allowing single-sided and cross ventilation (Chapter 9) and passive solar gain and daylighting. However, the overall energy balance will depend on the specific design. More complex atria with artificial heating may be of limited benefit.

Control at the building envelope

Control is necessary because solar gain, temperature and wind speed vary so much. Traditionally, occupants have been able to influence their environment and comfort by simple, easy-to-use, robust means and were then able to see, and almost immediately experience, the results of their actions. An example is the Victorian hospital window with a tall sash and a top hopper window operated by a crank whose control was at nurse height. As we develop more sophisticated systems it is important to try to keep these principles in mind. For example, opening inlets and outlets for natural ventilation systems that cannot be seen by the occupants introduce an element of uncertainty that poses problems for the psychological perception of comfort.

Controls can, of course, be manual or automatic or some combination of both. One probable development is more use of intelligent controls that allow occupants to override automatic systems for limited periods and then readjust according to conditions. For example, in a teaching space the lecturer might override the system to open the windows and allow more ventilation but the system would close the windows automatically at the end of each lecture period – or if it started to rain.

4.5 Internal layout

The internal layout joins the 'skin' to the 'body'. If there is only one space, or, in architectural slang, just one 'shed', the link is direct. In this case ventilation is straightforward, the noise is whatever comes through the skin and solar gains are relatively easy to deal with because they are entering a large volume.

As a space is subdivided, the situation alters. Partitions, particularly the full-height type, reduce the scope for natural ventilation, but ducted fresh air, probably with some mechanical assistance, can help overcome the problem. Opaque partitions interfere with views and reduce daylight penetration. However, partitions are likely to improve the acoustics by providing more privacy for conversations. Partitions and furniture also increase the admittance of a building – a figure of 1 W/K per m^2 of floor area has been estimated[26] – and this too can be an advantage.

Where loads are particularly high and daylighting requirements are low, as is often the case in lecture theatres – it could be advantageous to locate the space on the north side of the building, thus reducing solar gains.

Smoke and fire considerations will be mentioned in Chapter 9.

4.6 Form revisited

We can now consider form in greater detail and in particular look at two contrasting strategies of articulated versus compact forms.

An example exists in the UK where numerous government departments have been housed in collections of single-storey huts erected in the 1940s. With time these have been and continue to be replaced by multi-storey monoliths for economic reasons that include freeing land for more intensive development and lower building costs.

Environmentally, however, the arguments are less clear and somewhat contradictory. Compact shapes have low surface area/volume ratios and so are favoured where heat loss is a major issue – Inuits in their igloos know this. But there is less solar gain and daylighting available with the compact shapes and so, for example, energy consumption for artificial lighting will be higher. The significant potential for daylighting from the roof of single-storey buildings is one of their great advantages. Taller buildings will be more exposed to wind infiltration and will require energy for lifts.

Natural ventilation is much easier with articulated forms, which provide more possibilities for single-sided and cross ventilation (Chapter 9). Views out and contact with a natural or landscaped environment are also favoured by articulation.

Once again, the optimum strategy will depend on the application. Schools and office buildings require higher levels of lighting and higher ventilation rates than homes and so there is more call for articulated shapes with greater surface areas and more glazing. In a sense, the 'skin' services the building. One wants as much 'skin' as can usefully be employed but no more, because more 'skin' also means greater heat loss. We shall return to these questions in Chapters 8 and 9.

4.7 Two (more) models

Butterflies (Figure 4.9) are lightweight with powerful wings of large area compared with their bodies. Their sense organs vary from eyes for vision to antennae for smell. Butterflies are quick to respond to their environment. Elephants, on the other hand, are not lightweight. They keep a wary eye on their surroundings and are loathe to forget. If their environment changes, they also change – but only after a period of time.

Butterfly-type buildings will have highly responsive skins with a great deal of glass (and other materials yet to be developed) and will react quickly to changes in solar radiation, light and temperature, by altering their properties. They will also have ventilation openings that vary according to constantly changing needs. Parts of their envelopes will capture energy and generate power or heat directly just as some butterflies use the sun's radiant heat to warm themselves up so that they can get the full power of their muscles for flight.[27] Thermal mass will be incorporated but no more than necessary for vital building functions. Butterflies are 'high-tech'.

Elephant-like buildings have much more thermal mass. The building envelope is less critical because the mass compensates for the lack of a quick response. There are fewer openings and many of these may be manually controlled. Elephants are 'neo-vernacular'.

Of course, most buildings in the next few decades will be something in between as specific requirements for noise attenuation or heat disposal or increased views guide us to real designs.

A final word: as might be expected, the analogy does not support rigorous analysis. Butterflies are poikilothermic, i.e. their temperatures vary with those of their surroundings. Elephants are homeothermic and maintain a constant body temperature by internal means. Buildings with heating systems that work are homeothermic; those with systems designed by inexperienced engineers are poikilothermic.

Guidelines

1. Orientate the building to the south if possible.
2. Incorporate the right amount of thermal mass and high admittance surfaces into the building.
3. Increase the floor-to-ceiling heights in heavyweight buildings. Remember, the more height, the more light will enter.
4. Use glazing to allow solar gains and daylight but control at the building envelope to avoid overheating and glare.

4.9 Butterflies and elephants.

5. Incorporate a suitable degree of air tightness.
6. Specify windows and doors that are suitable for the degree of exposure and are detailed to reduce infiltration losses through them.
7. Insulate well to reduce heat loss.
8. Consider shutters or curtains to reduce night-time heat losses.
9. Decide if noise is a problem and, if so, how it will be attenuated.
10. Choose a compact or articulated form, or something intermediate according to suitability.
11. Use simple buffer spaces to reduce heat loss.

References

1. Milbank, N.O. and Harrington-Lynn, J. (1974) Thermal response and the admittance procedure. Current Paper 61/74. BRE, Garston.
2. Loudon, A.G. (1968) Summertime temperatures in buildings without air-conditioning. Current Paper 47/48. Building Research Station, Garston.
3. Anon. (1986) *CIBSE Guide A3: Thermal Properties of Building Structures*, CIBSE, London.
4. See reference 1, pp. 47–50.
5. See reference 1, p. 40.
6. See reference 3.
7. Anon. (1985) Energy efficiency in buildings. BS 8207, Appendix B, p. 13. British Standards Institution, London.
8. Anon. (1994) Minimising/avoidance of air-conditioning. Final Summary Report. Revision A. BRE Project EMC 32/91, Max Fordham & Partners, London.
9. Bordass, B., Entwisle, M. and Willis, S. (1994) Naturally ventilated and mixed-mode office buildings: opportunities and pitfalls. *CIBSE National Conference Proceedings, Brighton.*
10. Data from Anon. (1988) *Solar*, Monsanto, St Louis.
11. Anon. (1992) *Pilkington K Glass and Kappafloat*, Pilkington, St Helens.
12. Anon. (1992) Pilkington Data Sheets for Antisun, Reflectafloat, and Suncool Glass; and K Glass and Kappafloat. Pilkington, St Helens.
13. Pilkington Glass (1992) Private communication.
14. Anon. (1993) Double glazing for heat and sound insulation. BRE Digest 379. BRE, Garston.
15. Anon. (1979) *How windows save energy*, Pilkington, St Helens.
16. Littler, J. (1992) Smart glazing and its effect on design and energy, in *Energy Efficient Building: A Design Guide* (eds S. Roaf and M. Hancock), Blackwell, Oxford, pp. 101–128.
17. Ibid., p. 106.
18. Ibid., p. 111.
19. Twidell, J.W. and Johnstone, C. (1993) Improving low energy building design: experience from monitoring the world's largest building incorporating transparent insulation. *First International Conference, Environmental Engineering, De Montfort University, Leicester.*
20. Banham, R. (1969) *The Architecture of the Well-Tempered Environment*, The Architectural Press, London.
21. Anon. (1987) *Window Design: CIBSE Application Manual*, CIBSE, London.
22. Anon. (1981) Cited in CIBSE Building Energy Code, Part 2, CIBSE, London.
23. Anon. (1993) Double glazing for heat and sound insulation. BRE Digest 379. BRE, Garston.
24. Data from Airstream, Wokington, Berkshire.

25. See reference 16, p. 114.
26. M. Entwisle, Max Fordham & Partners (1994) Private communication.
27. Wigglesworth, V.B. (1964) *The Life of Insects*, The New American Library, New York.

Further reading

Anon. (1992) *Glass and Solar Control Performance of Blinds*, Pilkington, St Helens.

Anon. (1993) *Glass and Noise Control*, Pilkington, St Helens.

Baker, N.V. (Undated) *Energy and Environment in Non-Domestic Buildings*, Cambridge Architectural Research Ltd, Cambridge.

Hawkes, D. (1987) Energetic twosome. *Architects' Journal*, **185**(4), 40–7.

Saxon, R. (1994) *The Atrium Comes of Age*, Longman, Harlow.

Vale, B. and Vale, R. (1991) *Towards a Green Architecture*, RIBA, London.

CHAPTER 5

Site planning

5.1 Introduction

Around 30 B.C. Vitruvius wrote of the need to choose the most temperate regions of climate, since we have to 'seek healthiness in laying out the walls of the city' and went on to say that 'the divisions of the sites ... the broad streets and the alleys ... will be rightly laid out if the winds are carefully shut out from the alleys. For if the winds are cold they are unpleasant...'[1]

As Vitruvius knew, the site will have a marked effect on the functioning of the building and the building, in turn, will affect the site. Issues of particular concern are:

- site selection, microclimate and landscaping
- sunlight and solar gain
- daylight and views
- wind
- noise
- air quality.

5.2 Site selection, microclimate and landscaping

On a broad scale this book principally addresses issues for what has been termed the mid-European coastal climate as shown in Figure 5.1.

The European climate zones, which have fairly fuzzy boundaries, have been described as follows.[3]

1. Cold winters with low solar radiation and short days; mild summers.
2. Cool winters with low solar radiation; mild summers.
3. Cold winters with high radiation and longer days; hot summers.
4. Mild winters with high radiation and long days; hot summers.

One could add comments about the wind because of its importance for natural ventilation: the mid European coastal zone is characterized by a strong wind regime – at Heathrow airport in London, for example, a wind speed of 4 m/s is exceeded more than 50% of the time[4] (Figure A.3).

If we now consider sites and assume that a building site is required, consideration should be given to reusing existing ones if free, or to selecting locations that are least likely to be damaged environmentally if built upon. A checklist developed by the Building Research Establishment (BRE) for office buildings contains

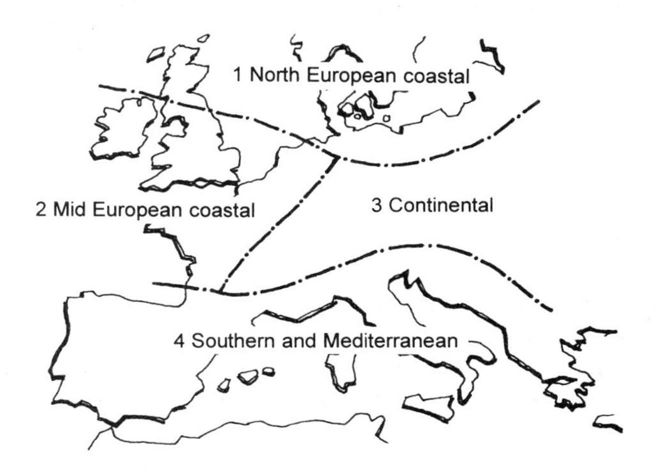

5.1 European climate zones.[2]

an examination of whether the site includes ecologically valuable features such as mature vegetation, ponds or streams and natural meadows and, if so, effectively attempts to direct developers towards other areas of lower ecological value.[5]

The next step is to try to create a combination of building and site which marries the aesthetic and physical environments. The elements of the physical environment include the site layout, the form of the building (in both plan and section), the materials used externally for the building and the landscaping (both hard and soft). Together, these factors will create a microclimate.

Traditionally, the primary concern of the microclimate in the UK has been to mitigate the effect of the 'cold, wind and wet of the relatively long cool season'.[6] With the current interest in natural ventilation (Chapter 9), a new concern can be added which is encouraging a microclimate that facilitates natural ventilation for cooling purposes during the relatively short warm season. We thus come to the first contradiction in microclimate design, namely that arranging the building so that it can be naturally cooled in summer may mean that ventilation heat losses in the winter are greater. Another major conflict, discussed in more detail below, is landscaping, which reduces wind speeds but may cause a loss of solar gain and natural light. Minor conflicts include:

- the use of dark surfaces near buildings to absorb solar radiation or light surfaces to reflect light into them, and
- water features that detrimentally dampen conditions in winter but are advantageous in helping produce a cool microclimate in summer.

Beneficial microclimates in the heating season are those that create warmer, dryer conditions and this can be done in a number of ways. The first is by taking advantage of solar gain (discussed principally in section 5.3), reducing the wind speed (discussed in section 5.5) and lessening the effect of rain. Rain can be dealt with by effective surface water drainage systems and this will favour hard, quicker drying surfaces. As always, however, a balance must be found because too extensive an area of hard surface causes problems with excessive run off. Thus, a combination of hard surface to take the water away from the buildings and soft landscaping to provide water storage capacity will be preferred.

Building form can also play an important role in controlling rain. For example, at Calthorpe Park School in Hampshire, 1200 mm deep roof overhangs were used both as sunscreens and shields to keep rain off the building fabric, thus protecting it and improving its insulating value.[7]

The net effect of a good microclimate is to reduce infiltration heat losses because wind speeds are lower, and fabric heat losses because the external temperature is somewhat higher.

5.3 Sunlight and solar gain

Buildings can, of course, be located completely or partially underground, as shown in Figure 5.2; as well as more traditionally above ground.

5.2 Ecology House at Stow, Massachusetts.[8]

Careful attention needs to be given to drainage, daylighting and ventilation of underground buildings, but fabric heat losses will normally be reduced because soil temperatures during the heating season are higher than air temperatures. Orientating partially underground buildings to the south will allow passive solar gain to contribute to the (reduced) space-heating requirements.

A variation on this theme is the single-storey building with extensive landscaped roof gardens and spacious courtyards that allow solar gain into south-facing spaces even in the winter. An example of this is discussed in Chapter 11.

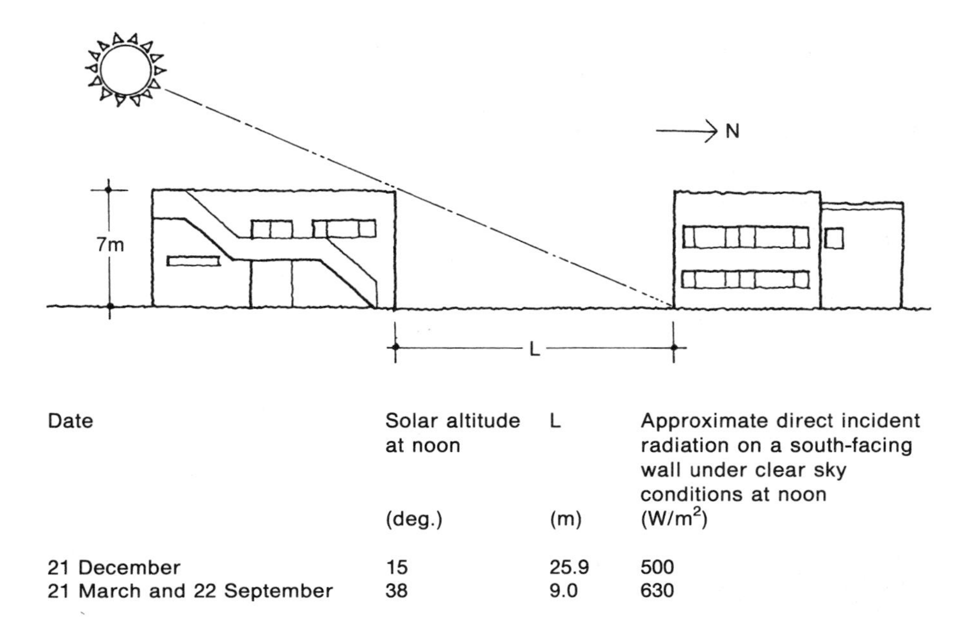

Date	Solar altitude at noon (deg.)	L (m)	Approximate direct incident radiation on a south-facing wall under clear sky conditions at noon (W/m^2)
21 December	15	25.9	500
21 March and 22 September	38	9.0	630

5.3 Spacings to achieve solar access at noon and radiation data for London.

Most buildings have been – and will continue to be – built above ground and for these the question is how to make the best possible use of solar gain in order to reduce energy consumption. Some use of solar gain is already made – it has been estimated that the Sun provides about 14% of the space-heating demands on average in UK homes.[9] Note that this is not all from direct solar radiation, but that an important contribution is made by diffuse solar radiation. It is easier in assessment techniques, however, to concentrate on direct radiation, and that is the approach followed below.

The Sun's position, of course, varies throughout the year – in London at a latitude of 51.5°N the solar elevation on 21 December is 15° at noon and this rises to about 62° at noon on 21 June.

Figure 5.3 shows the minimum north/south spacings required to give solar access at noon. Because before noon and after noon the solar altitude will be less, increasing these spacings will increase the number of hours of *solar access*. The direct radiation referred to in the figure is that portion of the solar radiation which comes directly through the atmosphere; sky diffuse radiation is that portion which is scattered back to Earth from the atmosphere.

Access to the Sun has both psychological and physiological effects that have always been appreciated. Figure 5.4 shows the magnificent sixteenth-century refectory of Fontevraud Abbey in France where solar gain through the large windows on the left of the figure was used to cheer the souls of the nuns who ate there.

A number of design tools exist for analysing solar access.[10,11] Computer techniques are being developed and will hopefully soon play an important role in determining the optimal spatial arrangement.

Work over the past 20 years has tended to concentrate on how to arrange large

5.4 Refectory at Fontevraud Abbey.

groups of houses to optimize use of the Sun's energy, and a typical layout is shown in Figure 5.5.

In such schemes a starting point is to get the road pattern right (roughly east–west); correct spacing should be dealt with at the same time. Common guidelines are to space houses in England more than twice their height apart and to orientate the long axis of the house within 45° of south.[13] If possible it is even better to orientate the house due south and to keep the sector 30° on each side free of obstructions, as shown in Figure 5.6.

For larger, more communal projects, the location of open spaces, gardens,

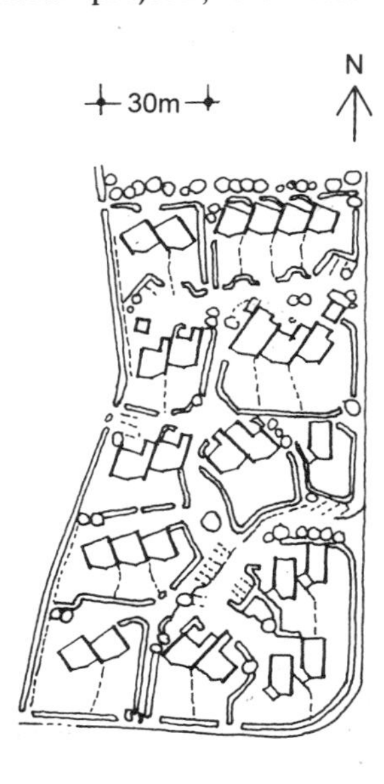

5.5 Housing at Angers, France.[12]

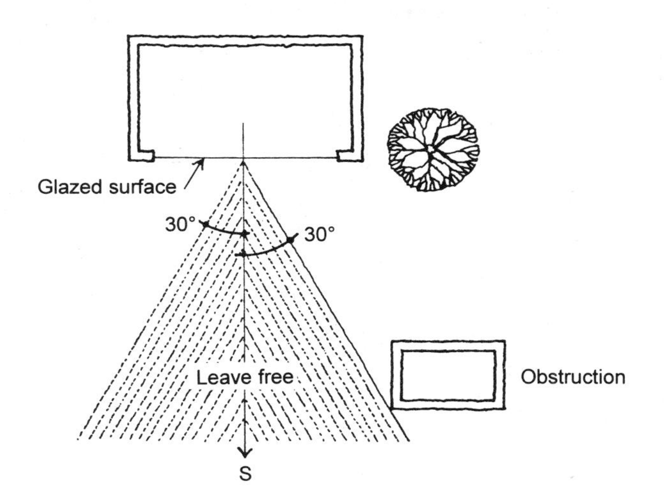

5.6 Orientation for passive solar gains in winter (based on Reference 14).

courtyards, garages and stores offers scope for facilitating solar access.

Trees and other vegetation have an important role to play in site layouts because of their amenity value and effect of tempering the wind (section 5.5). They can also provide some control of summer-time solar gain to avoid excessive temperatures at a cost of a winter-time loss of passive solar gain and a year-round loss of light (such trees effectively function as permanent, albeit seasonally variable, fixed external shades). Figure 5.7 shows a typical situation.

Any trees selected should, of course, be suited ecologically to the site. The designer can then consider, for deciduous trees, how long they are in leaf and how transparent they are to solar radiation, both in leaf and bare. Table 5.1 provides a selection of such data.

Thus, if we choose an elm for our tree in Figure 5.7 and locate it so that it will block out 85% of the Sun's radiation when in leaf in the summer (i.e. 15%

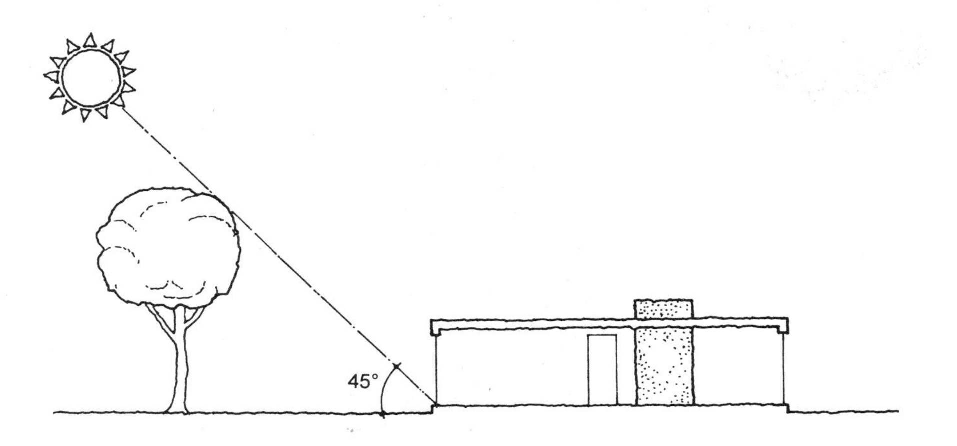

5.7 Effect of trees on solar access.

Table 5.1 Characteristics of common deciduous trees in the UK[15]

Botanical name	*Common name*	*Period of full leaf*	*Transparency (% radiation passing)*	
			Full leaf	*Bare branch*
Acer pseudoplatanus	Sycamore	May/August	25	65
Aesculus hippocastanum	Horse chestnut	Mid April/August	10	60
Betula pendula	European birch	May/August	20	60
Quercus roba	English oak	Mid May/mid October	20	70
Ulmus	Elm		15	65

Notes
1. Data is based on averages. Wide individual variations exist and so caution should be exercised.
2. Measurements are usually based on light but can be used for solar radiation also.

transparency), it will still block out 35% during the winter. Depending on the design, this can be a strong argument for less permanent solar shades.

Nonetheless, there tends to be a quite reasonable compromise between the amenity value of the trees and their functional role as windbreaks and solar screens. One approach is to locate deciduous trees to the south of southerly orientated buildings and to site lower evergreens to the north as a windbreak. The lower evergreens can, of course, also be used around the site for privacy and to the south as a windbreak (section 5.5).

5.4 Daylight and views

Daylighting of a space through a window is a function of the amount of sky the building can 'see' and, to a lesser extent, reflection from the surrounding surfaces. Assessing daylight access is somewhat similar to assessing solar access but differs in applying to surfaces at all orientations. Existing assessment techniques consider daylight availability, the effects of external obstructions and the reflectivity of external surfaces (Chapter 8 and Appendix C).

Daylighting guidelines exist for new developments. The BRE suggests that, as a first step, a check is made to see if there are obstructions within 25° of a reference line,[16] as shown in Figure 5.8.

If obstructions are at less than a 25° angle, the BRE advises that there will be potential for good daylighting in the interior, and an obstructing building that is 'too tall' but narrow may still permit good daylighting. Often site constraints, however, will not allow this criterion to be met. At the De Montfort Queens Building (Chapter 13), the spacing between the electrical laboratory wings was narrow (Figure 5.9) but by using white high-density panels as the external cladding the architect, nonetheless, achieved a light feeling in the courtyard and reasonable light levels in the interiors. It may be of value to consider an 'external' daylight factor in these cases. Point P receives about 40% of the light incident above the building at point A in overcast sky conditions.

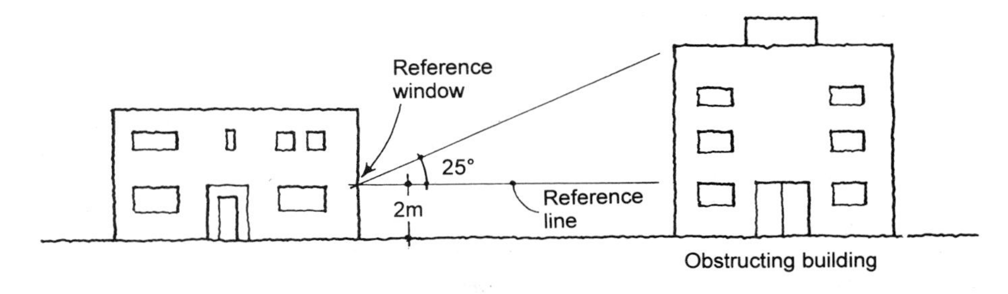

5.8 Angular criterion for spacing of building.

In even more constrained situations, as in many urban developments, the primary consideration is, in fact, ensuring that any new development does not affect its neighbours' 'right to light', and planning consent may depend on a successful solution to this problem.

We have seen how vegetation can have an adverse effect on solar gain in the winter, and since daylight is one part of solar radiation, it will, of course, be similarly reduced.

Just as a 'right to light' exists effectively for buildings, a right to a view ought to exist for people. Picasso used to say that he liked a view but preferred to sit with his back to it; most of us, however, would prefer some contact with the outside, whether this is to see changing sky conditions or panoramic scenes of cities. In numerous projects, from factories to restaurants to, more conventionally, schools, we have found that views out have been a major ingredient in the building's success.

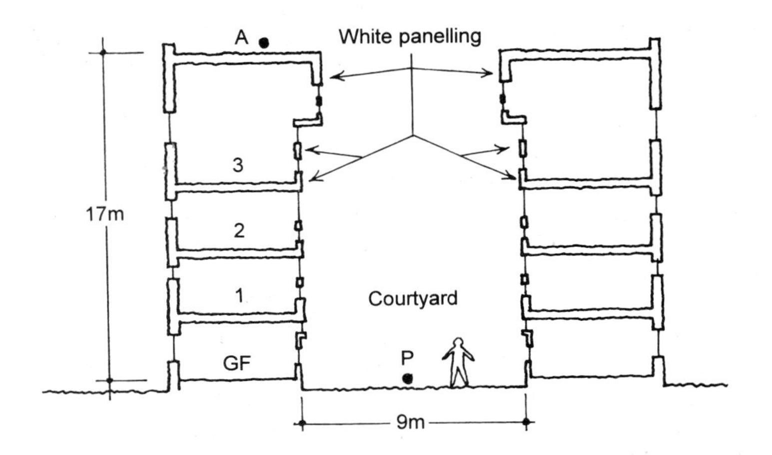

5.9 De Montfort University's Queens Building: electrical laboratory spacing (simplified).

5.5 Wind

A striking manifestation of wind forces on buildings is the flying buttresses of medieval cathedrals. As the cathedrals grew taller the architects and engineers found that wind forces (which are proportional to the square of the velocity of the wind) required a radical structural solution to maintain stability. The first flying buttresses were introduced at Notre-Dame de Paris in the twelfth century.[17]

Our concerns here are less critical but nonetheless important to the functioning of the building. The main aim in the heating season is to temper the winds around the site in order to:

- reduce the infiltration of external air into the building;
- increase the surface resistance of elements such as glazing, thus improving its thermal properties (eg the U-Value for double glazing at an 'exposed' site is about 3.2 W/m^2 K and at a 'sheltered' site about 2.8 W/m^2 K);
- reduce the wetting of the fabric by wind-driven rain, thus helping to maintain its insulation properties (both as a resistance to conductive heat transfer and by keeping it dry, thus stopping evaporative heat loss).

Obviously, reducing the wind speed on a site will also make it a more environmentally comfortable and enjoyable space for those who use it.

It is important that if the building is to be naturally ventilated any measures taken to improve the winter condition do not worsen the summer one. In principle, this should not prove too difficult; the main consideration will be ensuring that the paths to the air inlets are relatively free. Air outlets will normally be higher (Chapter 9) and should pose less of a problem. Wind is also useful in carrying away heat and pollutants from a site, and enough movement must be retained to ensure this.

Designers have one main way of tempering the winds and that is the use of windbreaks, which are likely to be of vegetation but can also include, for example, fences and other buildings. In laying out a site and incorporating windbreaks or shelterbelts care has to be taken to ensure that vegetation, in particular, does not significantly reduce passive solar gain. Figure 5.10 shows an idealized shelterbelt for protection from westerly winds.

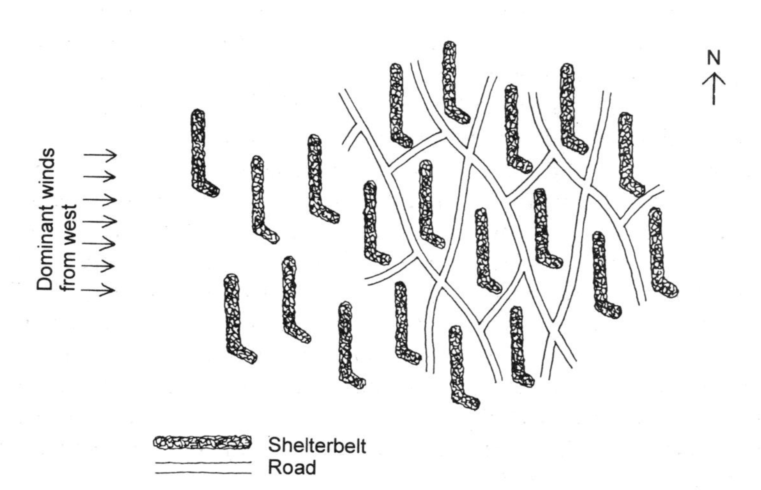

5.10 Idealized shelterbelt layout for protection from westerly winds.[18]

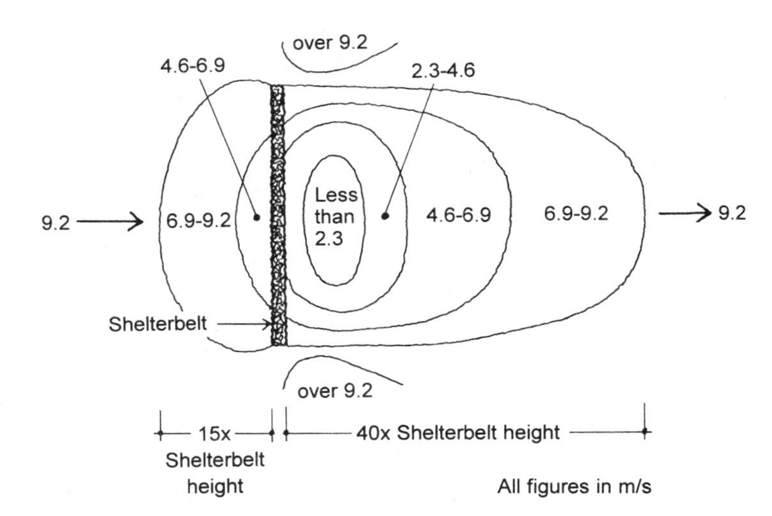

5.11 Reduction in wind due to a good shelterbelt.[20]

In the UK, the dominant winds are generally from the southwest and northwest but they vary significantly with place and time.[19] Designers should not rely on the wind being from a particular direction; however, they should try to determine the dominant wind direction in the summer and winter and draw the likely air movements on their site plans to help them visualize the flow patterns. Some data on wind is given in Appendix A.

The height of the shelterbelt and its permeability determine the area protected. Figure 5.11 shows an idealization of the effect of a windbreak.

Thus, if a shelterbelt is 10 m high, then up to a distance of about 120 m, the wind velocity will be below 50% of the reference wind velocity, which in Figure 5.11 is 9.2 m/s.

A number of rules of thumb exist. For example, a guide for energy efficiency in new housing suggests to reduce wind and to allow solar access to the site,[21] a shelterbelt should be located at a distance of three to four times its height from the homes to be protected. As always there is a balance to be struck among many factors. If, for example, the shelterbelts are far apart (to allow for solar gain), the increased wind speed may negate the effect of the increased solar input to the building. The goal is to optimize the overall performance of the microclimate of the site and buildings, and this remains more of an art than a science.

Designers can also influence the wind speed around a building through its form. The objective is to approximate forms that present the least resistance to the passage of the wind around them, thus reducing the disturbance to the wind pattern near the ground. It has been suggested that for normal, rectilinear buildings this implies a shape that is as near as practical to a pyramid.[22] Methods vary, from using hipped roofs rather than gable roofs for houses to stepping back the façades of multi-storey buildings. It must be said, however, that a pyramid has a greater surface area/volume ratio than a cube or a compact parallelopiped and so, inevitably, these techniques tend to increase surface area and raise the heat loss. Where the balance lies will depend, in the by now familiar way, on the site and on a variety of other factors.

Groupings of buildings can also be made less sensitive to the wind by using irregular patterns (Chicago became 'The Windy City' in part because of its regular street pattern, facilitating winds off Lake Michigan), keeping heights of buildings fairly uniform and creating courtyards. These and other techniques are described in References 23 and 24.

5.6 Noise

Careful arrangement of buildings and the use of shelterbelts can also improve the acoustic aspects of a site. Attenuation can vary from 1.5 to 30 dB per 100 m of shelterbelt, depending on the type of vegetation in the shelterbelt.[25]

5.7 Air quality

The air quality of a site can be improved by ensuring that winds can cleanse it and through the use of vegetation.

In photosynthesis plants absorb carbon dioxide and produce oxygen, and through transpiration they absorb water at the roots and release it into the air, principally at the leaves. Plants can also cleanse or filter the air when dust and pollutants adhere to their dry twigs or leaves (which are eventually washed by rain and impurities are deposited on the ground). Thus, highly planted zones will have a higher oxygen content, higher relative humidity and fewer pollutants and are likely to provide the right type of area from which to draw the supply air for a natural ventilation system.

Guidelines

1. Select a suitable site.
2. By siting of the building and the use of landscaping, develop a favourable microclimate with a suitable temperature, wind and relative humidity regime.
3. Orientate and space buildings to make use of passive solar gain and daylight.
4. Provide occupants with views out.
5. Use shelterbelts to temper the wind (draw the summer and winter wind patterns on a site plan).
6. Improve the noise climate on the site through grouping of buildings and the use of vegetation.
7. Remember that vegetation improves air quality.

References

1. Vitruvius. *The Ten Books on Architecture*, Book 1 (transl. F. Granger, 1931), Heinemann, London.
2. Baker, N. *et al.* (*ca.* 1992) *The LT Method: Version 1.2*, Commission of the European Communities.

3. Ibid.
4. Anon. (1986) *CIBSE Guide A2: Weather and Solar Data*, CIBSE, London.
5. Prior, J. (ed.) (1993) *BREEAM/New Offices: Version 1/93*. An Environmental Assessment for New Office Designs. BRE, Garston.
6. Anon. (1990) Climate and site development. Part 2: Influence of microclimate. BRE Digest 350. BRE, Garston.
7. Davies, C. (1985) Hants improvements. *Architectural Review*, **188**(1056), 22/2–29/2.
8. J.E. Barnard, Jr (1982) Private communication.
9. Anon. (1993) Low energy design for housing associations. *BRECSU Good Practice Guide* No. 79. BRE, Garston.
10. Anon. (1992) Lighting for buildings. BS 8206: 1992. British Standards Institution, London.
11. Ne'eman, E. and Light, W. (1975) Availability of sunshine. BRE Current Paper 75/75. BRE, Watford.
12. Dodd, J. (1989) Greenscape 2: Climate and form. *Architects' Journal*, **16**(189), 81–5.
13. See reference 9, p. 10.
14. Littlefair, P.J. (1992) Site layout for sunlight and solar gain. BRE Information Paper 4/92. BRE, Garston.
15. Anon. (1990) Climate and site development. Part 3: Improving microclimate through design. BRE Digest 350. BRE, Garston.
16. Littlefair, P.J. (1992) Site layout planning for daylight. BRE Information Paper 5/92. BRE, Garston.
17. Mark, R. (1990) *Light, Wind and Structure*, MIT Press, Cambridge.
18. See reference 15.
19. See reference 4.
20. Anon. (1964) *The Farmer's Weather*, Ministry of Agriculture, Fisheries and Food, Bulletin No. 165. HMSO, London.
21. Anon. (1993) Energy efficiency in new housing. *BRECSU Good Practice Guide* No. 79. BRE, Garston.
22. See reference 15.
23. See reference 15.
24. Dodd, J. (1989) Greenscape and tempering cold winds. *Architects' Journal*, **189**(18), 61–5.
25. Johnston, J. and Newton, J. (1993) *Building Green*, London Ecology Unit, London.

Further reading

Dodd, J. (1993) Landscaping to save energy. *Architects' Journal*, **198**(1), 42–5.

McHarg, I. (1971) *Design with Nature*, Doubleday, New York.

CHAPTER 6

Materials and construction

6.1 Introduction

'The subject of Material is clearly the foundation of Architecture', said William Morris in 1892[1] and now, a century later, with a far wider range of materials at the designer's disposal and more awareness of the environmental impact of materials, the statement has added significance. Materials affect structure, form, aesthetics, cost, method of construction and internal and external environments. This chapter examines basic criteria for their selection and provides data on those commonly used in buildings. It includes brief discussions of some construction issues and environmental assessment techniques.

6.2 Selection of materials

We should ask what criteria should be used when selecting materials before examining any specific ones. Relevant considerations (which complement the usual ones such as fitness for the purpose, cost, mechanical resistance, stability and safety) include impact on the natural environment and impact on health, with the two often being related. The impact on the natural environment includes ecological degradation due to extraction of raw materials, pollution from manufacturing processes, transportation effects, energy inputs into materials which affects CO_2 production and CFCs and HCFCs. Health issues range from how materials are extracted to the effects on the manufacturing workers producing the materials and to the internal environment that results from the materials selected. The major topics are discussed below but it should be remembered that the entire field is in a state of flux, and as more is learned about materials and the environment, conclusions will change. To cite but one example: in the 1970s the Cambridge University Autarkic House project developed a home intended to minimize energy use and supply the much reduced demand with solar and wind energy.[2] The intention was to use 700 mm of polyurethane for the thermal store insulation, and only later was it realized that the CFCs used in the manufacture of polyurethane were a major environmental hazard.

6.3 Environmental aspects of materials

The subject is, of course, enormous, and below we touch on only a number of key issues. Reference 3 provides additional information.

Sustainable development has been defined, none too precisely, as 'development which meets present needs without compromising the ability of future generations to achieve their needs and aspirations'.[4] How much of which materials should we be using? is an unanswered question. Should we prohibit the use of rare materials or should they be acceptable when 'absolutely necessary'?

An important question is: Where do the materials come from? For building materials areas of concern include land-take which exceeds 2000 hectares per year in the UK for aggregates (crushed rock, sand and gravel).[5] Few alternatives exist but some recycling is possible. Marine extraction of aggregates is eschewed by many as unnecessarily environmentally damaging.

Deforestation is another key issue but not a new one. England's forests have been reduced from about 1.8 million hectares in 1500[6] to about 1.0 million at present. Early uses of the timber included shipbuilding and fuel for iron-smelting and building. Environmental attention is now concentrated on more remote areas such as the rainforests. One approach – which has proved to be difficult in practice – is to ensure that any tropical hardwoods (such as mahogany, afrormosia, iroko and kapur) come from sustainable sources. The issue is a vexed one but guidance is available from a number of organizations including Friends of the Earth.[7] The concept of buying from sustainable sources is also being applied to hardwoods and softwoods from other areas.

Manufacturing and processing form another broad area of concern. Timber preservation, for example, is often essential for longevity but the chemicals used in the process need careful selection and handling. Concern about disposal affects many products, including plastics. PVC (polyvinyl chloride), for example, represents about 25% of total worldwide plastics production and is widely used in buildings for sheathing electric cables and for drains, cladding, floor coverings and window frames. Environmental concern has focused on recycling and burning of chlorides and release of pollutants.[8] The situation needs to be kept under review. A useful discussion of PVC in buildings, which covers such issues as fire safety and environmental effects, has been produced by MK Electrical Ltd.[9]

6.4 CFCs, HCFCs and HFCs and halons

Perhaps the most publicized and dangerous environmental issue is the depletion of the ozone layer mentioned in Chapter 2. Terminology in this area is important but unfortunately confusing. CFC stands for chlorofluorocarbon and refers to an organic molecule with chlorine and fluorine atoms. A measure of the damage caused to the ozone layer is a substance's ozone depletion potential (ODP). The ODP of the CFC known as refrigerant R11 is defined as 1.0 and other refrigerants are referred to it – the closer the ODP is to zero, the better the refrigerant is environmentally. In addition to their harmful effect on the ozone layer, refrigerants also contribute to the greenhouse effect, or global warming as we saw in Figure 2.5. Table 6.1 gives data for a number of refrigerants and halons. As can be seen from the table, most of these substances are much more harmful greenhouse gases than CO_2, which explains why, although the volume of such gases produced is relatively small, they account for about 10% of global warming.

Table 6.1 Characteristics of refrigerants and halons[10]

Substance	*Type*	*Formula*	*Montreal Protocol*	*Ozone depletion* potential *(R11 = 1)*[a]	*Global warming* potential[a] *(CO_2 = 1)*	*Flamm-ability*
R11	CFC	CCl_3F	Y	1	1500	No
R12	CFC	CCl_2F_2	Y	1	4500	No
R22	HCFC	$CHClF_2$	(N)	0.05	510	No
R113	CFC	CCl_2FCClF_2	Y	0.8	2100	No
R114	CFC	$CClF_2CClF_2$	Y	1.0	5500	No
R115	CFC	$CClF_2CF_3$	Y	0.6	7400	No
R123	HCFC	$CHCl_2CF_3$	(N)	0.02	29	No
R124	HCFC	$CHClFCF_3$	(N)	0.02	150	No
R125	HFC	CHF_2CF_3	N	0	860	No
R134a	HFC	CF_3CH_2F	N	0	420	No
R141b	HCFC	CH_3CCl_2F	N	0.08	150	Slight
R142b	HCFC	CH_3CClF_2	(N)	0.06	540	Slight
R152a	HFC	CH_3CHF_2	N	0	47	Moderate
R500		R12/R152a	Y[b]	0.74	3333	No
R502		R22/R115	Y[b]	0.33	4038	No
H1211	Halon	CF_2ClBr	Y	3.0	–[c]	No
H1301	Halon	CF_3Br	Y	10.0	5800	No
H2402	Halon	$C_2F_4Br_2$	Y	6.0	–[c]	No

[a]Global warming and ozone depletion potentials are per unit mass, and values are current best available estimates which may be subject to revision. Global-warming potentials relate to the long-term (500-year) warming potential. (N) in the Montreal Protocol column means that the substance is an HCFC and is expected to be phased out between 2020 and 2040 or earlier as alternatives are developed.
[b]R500 and R502 are implicitly included in the Montreal Protocol because they contain the restricted refrigerants R12 and R115.
[c] Value not yet measured.

The Montreal Protocol referred to in Table 6.1 is the 1987 multi-national agreement on reduced refrigerant and halon emissions into the atmosphere.

HCFCs are hydrochlorofluorocarbons. They contain chlorine but have lower atmospheric lifetimes than CFCs and are less damaging to the ozone layer, as indicated by their lower ODPs.

CFCs and HCFCs have both been commonly used and continue to be for the moment as refrigerants in air-conditioning systems, in commercial and domestic refrigerators (your home refrigerator is very likely to have the CFC R12) and in the manufacture of foamed thermal insulation materials. There is at present a shift away from CFCs (whose manufacture is being phased out) towards HCFCs and HFCs.

HFCs are hydrofluorocarbons. They contain no chlorine and have a negligible effect on the ozone layer (the ozone depletion potential of HFCs is estimated to be a thousandth of that of R11[11]) but they do contribute to global warming.

Because of this, environmental groups are searching for radical alternatives. One apparent success has been the Greenpeace refrigerator developed with the German company Foron, which uses no CFCs, HCFCs or HFCs. Instead, the refrigerant is about 20 g of propane and butane (deemed to be of minimal flammable risk because of the small quantity involved) and the insulation is blown using pentane.[12]

Halon is another imprecise term. In its broadest sense halon refers to all halogenated hydrocarbons and so can include, for example, CFCs. However, in its commonly used restricted sense it refers to halogenated hydrocarbons with bromine, which are used in fire-fighting systems. Halons extinguish fires by interfering with free radical chains. (Free radicals are highly reactive atoms or groups of atoms with unpaired electrons.)

Unfortunately, the properties that make halons useful in fighting fires also mean that ozone is destroyed in the atmosphere; thus, the ODPs of halons are very high (Table 6.1). There is a movement towards avoiding fixed gas-flooding fire-fighting systems altogether, but if this is not possible it is preferable to use CO_2 rather than halons. Thus, for the document store for RMC (Chapter 11), for example, CO_2 was specified. For hand-held systems CO_2 is often chosen but water spray, powders and foams are also available. Factors in selection include the cost and degree of potential damage to furnishings and equipment.

6.5 Materials and health

Materials extraction and product manufacture are critical health and safety points. It is worthwhile noting that our environmental and health problems are not, sadly, novel. The charming churches of Norfolk are often built in knapped flint (i.e. fragments derived from nodules of almost pure silica). The knappers often worked in conditions of poor ventilation in an atmosphere of fine dust which caused silicosis and the premature death of many of them.[13]

More recently, asbestos has been identified as a major hazard to health if fibres are inhaled. Asbestos is unlikely to be specified in new buildings; blue asbestos (crocidolite) and brown asbestos (amosite) are prohibited in the UK and white asbestos (chrysotile) is only permitted in certain formulations including asbestos–cement products but its disposal is a common problem when existing buildings are refurbished or demolished. This reminds us of the need to consider the full life cycle of any material or energy source.

Materials of high radioactivity should obviously be avoided because of the health hazard, but there are many other materials for which the danger is not necessarily as evident. These vary from products which release formaldehyde, those manufactured with or incorporating certain solvents, timber products treated with hazardous chemicals (for example, the insecticide HCH known as lindane) and composite materials incorporating certain resins. Each needs to be judged on its own merits.

Some paints traditionally have incorporated toxic metals such as cadmium (cadmium yellow was a favourite of the Impressionist painter Monet in his later works) and lead. Outdoor paints often incorporated lead for added weather protection. The main health risk of lead-based paints is their ingestion by children.[14] As effective lead-free paints are widely available, their use should be encouraged; this has been recognized and legislation will soon make lead-free paint mandatory.

Health in the workplace is a major environmental issue which reflects the enormous amount of time we spend in these relatively sealed areas. Indoor air quality (mentioned briefly in Chapter 2) is part of this. Another area of concern that needs to be monitored is the effect on health of electromagnetic fields due to electrical distribution systems and electrical equipment, including such mundane devices as hairdryers.[15]

6.6 Materials and energy

On 3 February 1695 at Versailles inside the Hall of Mirrors it is said that the temperature dropped to the point where wine and water froze in the glasses. It was an exceptionally cold year, but even in more clement times the heating system – consisting of two open fireplaces – consumed and furnished only relatively small amounts of energy. The energy that had gone into the splendid stone and decorations of the Hall, however, was significant. Both the running energy and initial energy were derived mainly from renewable sources, in particular wood, water and wind power; coal was available and had, for example, been used at least since the twelfth century for lime production,[16] but its cost limited its employment.

By contrast, energy inputs for running buildings now tend to be much greater than the initial energy inputs. Initial energy, or, more precisely, embodied energy, has been defined as the energy used to (a) win raw materials, (b) convert them to construction materials, products or components, (c) transport the raw materials, intermediate and final products; and (d) build them into structures.[17] The figures do not include maintenance, reuse or final disposal. Determination of the embodied energy is a field fraught with uncertainty for a number of reasons, including the difficulty of standardizing data and incomplete knowledge of the fuel mix used in production. The field is also rife with debate as manufacturers stake rival claims to lower embodied energy and, thus, lower 'embodied' CO_2 production. Furthermore, it is an area that changes as manufacturing processes evolve.

In the UK, approximately 5–6% of the total energy consumption is embodied in construction materials[18] compared with about 50% used in buildings for space heating and cooking, water heating, lighting and power (Chapter 3).

For new office buildings as a whole, the embodied energy ranges from 3.5 to 7.5 GJ/m^2 of floor area whereas energy in use amounts to between 0.5 and 2.2 GJ/m^2 yr; typically, the initial embodied energy of an office is equivalent to about five years of energy in use, or about 7% of the total energy used over the lifetime of the building.[19] Obviously, as buildings become more energy efficient in use, the embodied energy will become relatively more important and, similarly, the relative energy involved in demolition and the importance of recycling materials will increase. At present, however, the greatest energy savings are to be obtained by reducing energy consumption in use.

In this field of embodied energy it is useful to try to find a position from which to take an overall view. Table 6.2 gives broad worldwide and UK comparative energy requirements for major building materials.

The table must be used cautiously. As is evident, there are major variations,

Table 6.2 Broad comparative energy requirements of building materials

Material	*Primary energy requirement (GJ/tonne)*		
	Worldwide[20]	*UK*[a21]	*UK*[22]
Very-high-energy			
Aluminium	200–250		97
Plastics	50–100		162
Copper	100+		54
Stainless steel	100+	75[a]	
High-energy			
Steel	30–60	50	48
Lead, zinc	25+		
Glass	12–25		33
Cement	5–8		8
Plasterboard	8–10		3
Medium-energy			
Lime	3–5		
Clay bricks and tiles	2–7	2	3
Gypsum plaster	1–4		
Concrete:			
In situ	0.8–1.5		1.2
Blocks	0.8–3.5		
Precast	1.5–8		
Sand–lime bricks	0.8–1.2		
Timber	0.1–5		0.7[b]
Low-energy			
Sand, aggregate	<0.5		0.1
Flyash, volcanic ash	<0.5		
Soil	<0.5		

[a]More complete data is available in the reference cited.
[b]Local air dried.

although the broad classification of energy bands seems about right. Generally, energy inputs will depend on a country's fuel mix, and the source of a country's materials will affect the figures. For example, timber varies depending on whether or not it is grown locally. Most softwood in the UK is imported and so the embodied energy includes a significant transportation component. Energy inputs into metals such as copper and aluminium can vary widely according to whether the source is from the ore or from recycled material.

If we now look at the main structural materials used in building, say, a typical detached house, we find that the bulk of the building relies on quite a small number of materials and that the embodied energy requirement of very-high-energy and high-energy materials is not a great percentage of the total. Figure 6.1 shows typical embodied energy inputs.

Obviously, the materials and the relative contributions of the various components will vary with building type and particular buildings. Reference 24

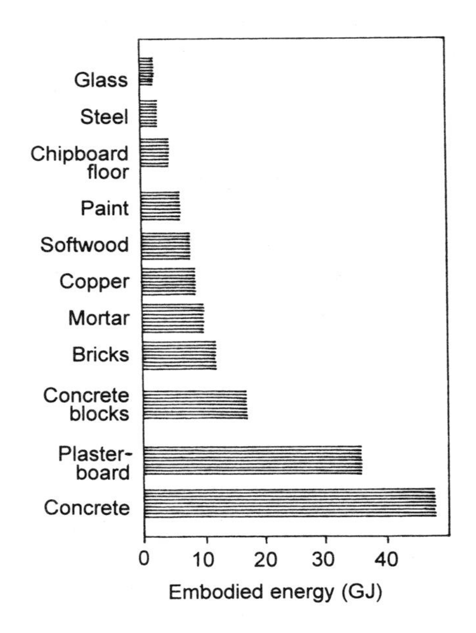

6.1 Approximate embodied energy inputs for a detached house (based on Reference 23).

gives an analysis of energy inputs into major building materials (glazing, finishes and services are all excluded) for local authority housing types, varying from a two-storey house to a nine-storey block of flats. In the former, steel reinforcement accounts for 1% of the total of components analysed and in the latter it is 40%!

The embodied energy in the services systems – mechanical, electric, above-ground soil and waste, rainwater disposal and drainage – obviously varies with the building type but services might account for approximately 5–10% of the embodied energy.

An analysis of the BRE Low Energy Office indicated that the gas-heating system accounted for about 220 MJ/m^2 of floor area, followed by ventilation at 120, lights at 60 and power at 20.[25] It is interesting to note that generally the cost of a component is a good indication of embodied energy. Services, however, are an exception having a higher proportional cost.[26] This might be due to the intricacy and the greater amount of thought that goes into, say, a boiler or air-handling unit than a concrete slab.

Environmental concerns have meant that the carbon dioxide production associated with energy use has received increasing attention. Carbon dioxide emissions resulting from both the embodied energy and the energy consumed in use over the building's life cycle have been studied for the BRE Low Energy Office referred to above[27] and findings are discussed in Chapter 9.

There are a number of ways in which to reduce the embodied energy and CO_2 production of buildings. The first is to select lower energy materials, paying attention to possible reduced energy substitutes for traditional solutions. A second is to design for longevity. This has a number of aspects, an important one of which is to design buildings of excellence that are acclaimed publicly as they are likely to be well maintained and be in existence for many years. Other aspects include high-quality, durable materials and design solutions that reduce the need for refurbishment. A third aspect is economic use of materials and designing to reduce waste. Recycling of materials is a final and very important aspect. Some materials such as lead have traditionally been recycled; indeed, as long ago as 1775 Samuel Johnson castigated (unjustly, it appears) the clergy of Lichfield for selling the cathedral's lead roof.[28] More recently, efforts have been made to increase the recycling of materials such as concrete and plastics. Theoretical studies are multiplying and include one that has considered the production of bricks from sewage sludge.[29]

6.7 Key materials

The selection of key materials and components will depend on a wide variety of factors: noise transmission, structural spans, fire considerations and cost are just a few of the constraints that guide the building process. In the sections below a brief look is taken at a number of key elements. The information needs to be examined critically and incorporated cautiously into any design. Interactions are numerous: for example, spaces with high exposed ceilings of concrete incorporate a fair amount of energy for a given floor area and the volumetric cost of the space is high; however, the increased mass may eliminate any need for air conditioning and the increased height can allow greater use of daylighting. Victorian era schools of heavyweight construction (little else was available) had high ceilings and large glazed areas to maximize daylight since artificial lighting was rare and expensive. Heating was minimal because it was expensive and pupils and staff were deemed to be robust. The solution had a high embodied energy cost but a low running energy requirement.

Structural materials

Structures obviously need stiffness and a common measure for this is the somewhat surprisingly named elastic modulus, E. Table 6.3 relates stiffness and energy – a higher value of E means a stiffer material.

Table 6.3 Energy requirements and stiffness[30]

Material	*Elastic modulus, E (MN/m^2)*	*Density (kg/m^3)*	*Energy (kJ/kg)*	*Energy cost of one unit of E*
Timber (sawn)	11 000	500	1 170	53
Mass concrete	14 000	2 400	720	124
Brick	30 000	1 800	2 800	167
Reinforced concrete	27 000	2 400	8 300	738
Steel	210 000	7 800	43 000	1 598
Aluminium	70 000	2 700	238 000	9 180

The lower energy cost for timber explains in part why there is a growing recognition that wood is an energy-efficient material. The Timber Research and Development Association (TRADA) has examined different methods of constructing a three-bedroomed detached single family house and concluded that a standard timber-framed wall requires 7450 kWh whereas a lightweight concrete block wall requires 12 816 kWh, or 1.7 times as much energy.[31] The figures are quoted cautiously, since it is not always evident that manufacturers – or researchers – make true comparisons. For example, timber elements may not have the same acoustic performance as concrete.

For large buildings, common options are structural masonry, steel frame and concrete frame. In an analysis prepared for Linacre College, Oxford, it was found that 'structural masonry and timber floors rather than frames and concrete floors are to be used where possible'.[32] For Linacre itself, however, precautions for partial collapse led to two-way spanning concrete floors.

Floors, walls and roofs

Again, for a detached house, TRADA has found that a timber ground floor requires 2669 kWh compared to a concrete slab on the ground at 6922 kWh, i.e. 2.6 times more energy; for intermediate floors, timber required 2947 kWh compared to concrete at 8312 kWh.[33] However, another study of housing found that a concrete slab used only 1.07 times the energy of a complete construction of suspended timber floor plus bricks and block.[34] Timber floors have less embodied energy than concrete, but the degree of difference is a matter of debate.

External wall systems vary from lightweight cladding to traditional masonry construction. Some lightweight wall constructions embody relatively low amounts of energy but are not always aesthetically suitable. TRADA claims that, for the same detached house referred to above, a timber-framed wall requires 7450 kWh compared to a lightweight concrete block wall at 12 816 kWh, i.e. 1.7 times more energy.[35] Solid concrete block walls have been found to require less energy than solid brickwork.[36]

For internal walls, stud partitions require about 6.4 kWh/m^2 compared with 7.7 kWh/m^2 for blockwork with a plaster skim;[37] however, consideration must be given to structural strength, thermal capacity, acoustics and longevity.

Roofs

The study for Linacre College referred to above found that pitched timber and tile roofs required less embodied energy than steel-framed roofs and that these in turn were better than flat concrete with asphalt.[38] Again there is a recurring theme that the use of timber is efficient in its use of embodied energy.

Insulation

Insulation levels have been a prime area of development in energy conservation for several decades. The flow of heat through a wall depends on the insulation

Table 6.4 Insulation data

Material	Thermal conductivity (W/m K)	Density[a] (kg/m^3)	Thermal resistivity (m K/W)	Embodied energy (kWh/m^3)	Embodied energy/thermal resistivity (kWh.W/m^4 K)
1. Expanded polystyrene slab	0.035	25	28.6	1125[d]	39.3
2. Extruded polystyrene	0.030[b]	38[b]			
3. Glass fibre quilt	0.040	12			
4. Glass fibre slab	0.035	25			
5. Mineral fibre slab	0.035	30	28.6	230[d]	8.0
6. Phenolic foam	0.020	30			
7. Polyurethane board	0.025	30			
8. Urea formaldehyde foam	0.040	10			
9. Cellulose fibre	0.035[c]	25[c]	28.6	133[d]	4.7

[a]Data from Reference 42 unless otherwise noted.
[b]Reference 43.
[c]Reference 44.
[d]Reference 45. *NB* Values in the literature vary greatly, and caution must be exercised. The values cited are all taken from Reference 45 for consistency; however, they should be used indicatively only at this stage because there is no strong agreement among figures from different references. For example, Reference 46 gives a value of 59.2 kWh/m^3 for mineral fibre insulation.

level and Appendix B gives a sample calculation. However, the insulation level is only one factor in the energy consumption of a building and often the correlation between insulation levels and energy consumption is not great. Indeed, in practice it is often difficult to find correlations because so many factors are involved in the energy consumption of buildings, including how the buildings are used by the occupants. For example, recent research on single-building primary schools in Scotland showed that it was difficult to draw any conclusions about the effect on energy consumption of insulation standards, shape, orientation, exposure to weather or type of heating controls.[39] The factors that were found to affect energy consumption were floor area, building volume and the number of occupants.

If the insulation meets current Building Regulations then the heat flow through the envelope may be only a small proportion of the total energy required by the building. At this point, it may be cost-effective to look at alternative methods of saving energy, such as ventilation control.

Nonetheless, the basic questions remain: Where should the insulation be placed? What should it be? And how much is needed? Masonry construction with a cavity continues to dominate UK structures. At the International Headquarters of RMC (Chapter 11), completed in 1989, the wall construction of 150 mm lightweight block, a 50 mm cavity and 100 mm external brickwork, gave a U-value of 0.54 W/m^2 K. At the De Montfort University Queens Building (Chapter 13), completed in September 1993, 190 mm block (used for structural reasons), 100 mm of Rockwool cavity batts completely filling the cavity and 100 mm external brick as shown in Figure B.1, gave a U-value of 0.30 W/m^2 K (Appendix B).

This reflects a continuing trend towards lower U-values and perhaps also a greater confidence among designers that the cavity can be completely filled without risking moisture transfer from outside to inside.

Even lower U-values can be and have been achieved in a number of ways, but one of the easiest is by increasing the cavity width and filling it almost completely. For example, a construction of 100 mm external brick, 165 mm cavity with 150 mm of mineral wool and 100 mm internal brick with 12 mm dense plaster achieved a U-value of about 0.18 W/m^2 K;[40] special permission was required from the Building Inspector for this width of cavity.

The principal insulation materials currently available and their characteristics are shown in Table 6.4, and a useful overview of insulation materials can be found in Reference 41. The last column in the table is, of course, comparable to the final column of Table 6.3.

Debate on the choice of insulation materials tends to be related to environmental considerations and buildability. If we first consider the environmental aspects (and keep in mind that it can be misleading to talk about an element out of context, i.e. insulation without considering the entire, say, wall construction) a principal concern for plastic insulants is ensuring that neither CFCs nor HCFCs are used as blowing agents during their manufacture. The situation is in flux and so individual manufacturers must be contacted on their products. Expanded polystyrene, for example, now tends to be free of CFCs and HCFCs whereas this is still not always true for extruded polystyrene.

For fibrous materials (such as glass and mineral fibres), concern has been expressed about a possible increased risk of lung cancer or non-malignant diseases such as bronchitis. The issue is highly technical and is related to fibre diameter, length and chemical composition; to give a quantitative tone to the discussion, Rockwools, for example, have fibre diameters of 4–5 μm.[47] Recent reports, however, indicate that there is not an increased health risk associated with the use of these materials (see Reference 48 for a summary discussion of these issues). This is a subject (among many in the environmental field) that should be kept under review.

With regard to embodied energy, it can be seen from Table 6.4 that there is a wide range of values. Insulation materials derived from mineral fibres tend to require less embodied energy than a number of others; similarly, they have lower CO_2 emissions. Researchers have noted that there are large discrepancies in the literature between embodied energy for similar materials and that careful consideration of the validity of data is required.[49] There is general agreement, however, that the use of insulation saves many times its embodied energy – manufacturers' figures range from the hundreds (200 for, say, expanded polystyrene) up to a thousand times (for, say, glass fibre). CO_2 emissions follow a similar pattern of major savings.

Turning to buildability, some architects are concerned about friability of fibrous materials and it is important to ensure that COSHH Regulations are met during installation of such products. Many, however, find that Rockwool products suit their designs very well and specify it regularly. Housebuilders are sometimes said to prefer extruded polystyrene foams because of the ability to lap boards and maintain a clear cavity and good water resistance.

Glazing

Glazing depends on high-energy materials but provides the priceless possibility of views and the more easily priced potential for passive solar gain and daylighting.

Embodied energy figures tend to increase with complexity of design. Wooden-framed single glazing provides a baseline and as one moves towards argon-filled, sealed double-glazed low-emissivity coating units in aluminium frames with thermal breaks the energy cost rises. Life cycle analysis indicates that the reduced energy in use of high performance glazing systems compared with single glazing in wood frames is recovered, very broadly, in 5 to 20 years, and that wooden frames appear to offer significant potential for reduced embodied energy.[50]

The monetary cost of high-performance glazing systems is another issue. The economics of double glazing is debatable but it has nonetheless become an almost standard feature of new buildings for a number of reasons including environmental ones. Double glazing with low emissivity coatings is probably not cost effective in most applications.

Services

Underground drainage often consists of clay pipes with plastic connectors; above ground soil and waste pipework has moved (quickly) away from traditional cast iron to plastics. Rainwater disposal components tend to be plastic or made of metals such as aluminium. Thus all of these systems have energy implications. However, heating, ventilating, air-conditioning, power and lighting are the main areas of interest.

The best way to reduce the environmental impact of services is to reduce the need for them. Nonetheless, there tends to be an unavoidable minimum input (heating in the depths of winter, lighting at night) and meeting this merits close attention.

A comparison of gas and electric heating systems is given in Chapter 9, as is a discussion of mechanical ventilation systems and heat recovery.

Pipes are usually in steel or copper (lead pipes which are to be avoided, were common in the past but are not normally used in new installations) and ductwork is mainly in galvanized steel; plastic substitutes for both are becoming more common. All have energy and material implications. Pipe insulants selected should be CFC and preferably HCFC free.

Power and lighting cables tend to be in copper with PVC insulation. As discussed previously, the PVC use needs to be kept under review.

Lighting

Environmental concerns focus on the materials in lighting systems and the effects of providing power to the lights. In the UK 60 000 000 failed fluorescent tubes are discarded annually. Each tube contains about 20 mg of poisonous mercury and so 1200 tonnes are disposed of each year along with significant quantities of cadmium, lead, copper, tungsten and a number of other pollutants.[51] The need for safe disposal of lamps is therefore becoming increasingly recognized.

Embodied energy (and CO_2 emissions) associated with lighting systems are low compared with consumption in use.[52] The luminaires (or, less technically, the light fittings – see section 8.3) account for over 80% of the embodied energy in the system and there is scope for lowering the energy content by reduced use of plastics.

Energy in use is, of course, very high (in the BRE low energy office only 14 months of use equals the embodied energy[53]). Generating this energy produces CO_2, as we have noted, and a variety of other pollutants: for example, electricity from coal-fired power stations results in emissions of mercury and other trace elements to the air. Thus, even though tungsten lamps do not contain mercury, their use in the UK with its mix of generating stations, including coal-fired, results in emissions of this metal. A quantitative life cycle analysis of major lamp types is to be found in Reference 54. The general conclusion is that light sources with higher luminous efficacy (Table 8.4) are more environmentally friendly.

As is common, there are conflicting claims by manufacturers and consultants about which lighting systems are the most environmentally friendly.

Finishes

The primary concerns about finishes have already been mentioned and fall into the two major categories of environment and health.

For floors, denser timber is more hard wearing but, unfortunately, it often comes from tropical rainforests. Again, the recommendation is to ensure that all woods used are from sustainable sources. Where possible, varnishes and paints should be based on water, plant oils or resins and those with fungicides, arsenics, harmful solvents and lead should be avoided.

Floor coverings, if required, should be from renewable materials where possible and, ideally, combine durability with a low embodied energy content. Coverings mainly based on renewable sources include linoleum and cork. Other coverings are often plastics based and may incorporate resins.

For walls and ceilings the first question is whether a finish is necessary. It may be possible to leave brickwork (as at De Montfort's Queens Building (Chapter 13)) or concrete ceilings (as both at De Montfort and RMC (Chapter 11)) free; this can also have thermal advantages in reducing peak summer temperatures.

The disadvantages of plasters and renders are that they are from non-renewable resources, require significant energy inputs and are not readily recycled.

Paints for walls and ceilings should meet the same criteria as paints for other surfaces.

6.8 Construction

As lack of space limits discussion of construction issues, we shall only mention a few points.

As insulation levels increase, construction methods change: for example, increased cavity wall insulation is likely to increase cavity width and change such details as the lengths of wall ties. Similar effects can result if less efficient insula-

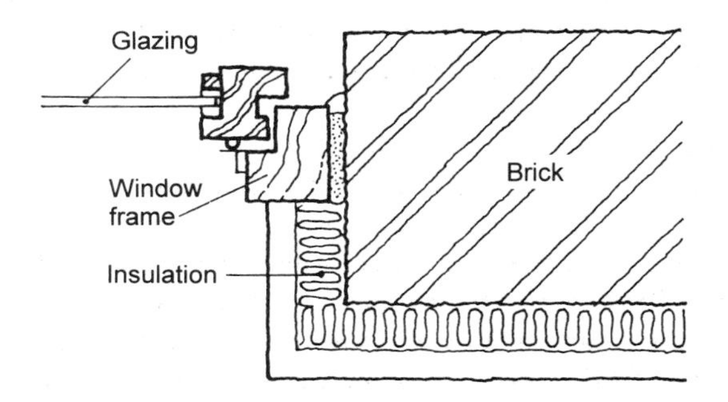

6.2 Avoidance of cold bridging at a window rebate – plan view.[57]

tion materials replace those blown with CFCs and HCFCs. These issues are noted in Reference 55.

As insulation is improved, the heat loss through the fabric becomes relatively less important and more attention needs to be paid to ventilation and the concomitant issue of possible condensation.

In our more tightly sealed buildings, condensation can occur in corners, behind doors, on walls and in roofs. Experience has shown that simply providing openings for air is not necessarily sufficient and that air must be able to circulate and take the moisture away. The greatest danger in roof voids occurs in cold, still conditions. The issues are highly technical and involve the performance specification of vapour checks. The main points of this aspect are introduced in Reference 56.

Similar problems can occur in refurbishment. Internal insulation of certain concrete system-built flats can result in condensation due to the high water vapour-resistance of the outer leaf. To reduce this risk, cavities can be vented by holes drilled through the outer leaf at high and low level.

To produce energy efficient buildings, cold bridges (local poorly insulated paths from inside to outside) need to be avoided. Figure 6.2 shows a simple example where the insulation has been carried around to a window rebate.

It is also important to reduce infiltration losses by filling in cracks in walls and ceilings, by sealing holes around pipes, drains and cables and by sealing around window frames with mastic.

On site, inspection is needed to ensure that the insulation and sealing details that have been drawn are achieved.

6.9 Evaluation methods

There is a saying that 'comparisons are invidious', and in the environmental field one can add, 'difficult and contentious'. If we first consider materials, current approaches include life-cycle analysis and eco-labelling. Life-cycle analyses examine the 'cradle-to-grave' or origin to disposal impacts of materials or products. The analysis of energy used is just one aspect (but perhaps the most easily quantifiable) of such studies which, ideally, should incorporate all environmental effects, in the broadest sense of the term.

Eco-labelling, or environmental labelling, is an attempt to arrive at agreed measures of the environmental suitability of products, which can range from light bulbs to hairsprays to washing machines. The difficulty of comparison is illustrated by the BSRIA Environmental Code of Practice which effectively grades approximately 30 types of material, ranging from timber to lime mortar, by assigning a value of 0–3 to 16 equally weighted categories,[58] including natural occurrence, radioactivity, acoustic properties and smell. Timber and beeswax are among the highest ratings and concrete reinforced with steel among the lowest, but the procedure is fraught with difficulties, such as varying suitabilities for the purpose, and also expounds some subjective judgements (timber, for example, is described as having very desirable thermal properties).

It would appear preferable to concentrate on major, well-defined issues – for example, energy inputs and savings for insulation materials, energy use of washing machines, NOx emissions of boilers (Chapter 9) – and to broaden the analysis from a sound base.

For buildings as a whole there are a number of schemes and, in the UK, BREEAM (Building Research Establishment Environmental Assessment Method) provides, for a number of building types, both a useful discussion of the major issues and a ready checklist. As discussed above, weighting criteria remain a problem with individual products. For new homes, one credit is given if all insulants have zero ozone depletion potentials and another is given if there are four containers for household waste to encourage recycling.[59] Are these comparable? Are they of approximately equal concern? What is certain is that the issues require widespread discussion and that there is a need for objective data.

Guidelines

1. Use materials with minimal health and safety risks over their life cycle. The use of hazardous materials such as asbestos should be avoided.
2. Avoid CFCs. Use HCFCs only when unavoidable.
3. Investigate the impact of extraction of source materials, pollution associated with manufacturing and disposal and the possibility of recycling materials.
4. Promote sustainability.
5. Embodied energy is important but the greatest savings are to be made from reducing energy in use.
6. Use energy efficient materials. This tends to mean more wood. Restrict the use of plastics and metal to situations where they are indispensable, e.g. copper and aluminium for cables, or where there are significant advantages in weight, strength or durability.
7. Insulate the building fabric well and use high-efficiency glazing systems.
8. CO_2 and other emissions associated with materials need to be considered.
9. The energy and materials implications of services systems have not received sufficient attention. Try to find out more about what will be used in your buildings.

10. The best drawn design can be ruined on site. Attention to the construction process is vital.
11. Evaluation methods are controversial but are often useful for purposes of comparison and as checklists.
12. Materials, and our knowledge of them, are constantly changing – keep abreast of recent major developments when specifying.

References

1. Cited in Clifton-Taylor, A. (1972) *The Pattern of English Building*, Faber & Faber, London, p. 6.
2. Littler, J. and Thomas, R. (1979) Thermal storage in the autarkic house. *Transactions of the Martin Centre for Architectural Studies*, University of Cambridge, Cambridge.
3. Anon. (1993) *The Green Construction Handbook*, JT Design Build, Bristol.
4. Cited in Elkin, T., McLaren, D. and William, M. (1991) *Reviving the City*, Friends of the Earth, London.
5. Ibid., p. 33.
6. See reference 1, p. 294.
7. Anon. (1993) *Timber: Types and Sources*, Friends of the Earth, London.
8. Alexander, D. (1993) War on PVC. *New Builder*, 25 June, p. 3.
9. Anon. (Undated – approximately 1992) *PVC in Buildings*, MK Electric Ltd, London.
10. Anon. (1991) CFCs and buildings. BRE Digest 358. BRE, Garston.
11. Baggott, J. (1994). HCFCs declared 'safe' for ozone layer. *New Scientist*, **141**(1911), 15.
12. Anon. (Undated – estimated as 1994) *Saving the Ozone Layer with Greenfreeze*, Greenpeace, London.
13. Mason, H.J. (1978) *Flint – The Versatile Stone*, Providence Press, Ely, p. 25.
14. Crisp, V.H.C., Doggart, J. and Attenborough, M. (1991) BREEAM 2/91. An environmental assessment for new superstores and supermarkets, BRE, Garston.
15. Maclaine, D. (1994) Fields of doubt. *Electrical Review*, **227**(10), 28–30.
16. Gimpel, J. (1975) *La Révolution Industrielle du Moyen Âge*, Seuil, Paris.
17. Howard, N. and Sutcliffe, H. (1993) Embodied energy. The significance of fitting-out offices. BRE 180/22/9, BRE, Garston.
18. Ibid., p. 1.
19. Ibid.
20. Spence, R. (1991) *Energy for Building*, United Nations Centre for Human Settlement, Nairobi.
21. Howard, N. (1991) Energy in balance. *Building Services*, **13**(5), 36–8. (*NB*: Figures in this article should be regarded as approximate as the database was under development at the time.)
22. Halliday, S.P. (1994) Environmental code of practice for buildings and their services. BSRIA, Bracknell.
23. See reference 21, p. 36.
24. Gartner, E.M. and Smith, M.A. (1976) Energy costs of house construction. Current Paper 47/76. BRE, Garston.
25. Butler, D. and Howard, N. (1992) From the cradle to the grave. *Building Services*, **14**(11), 49–51.
26. See reference 17, p. 2.
27. See reference 25.
28. See reference 1, p. 380.
29. Bryan, E. (1993) Residuals as resources. *Proceedings of the First International Conference on Environmental Engineering, De Montfort University, Leicester.*
30. See reference 20, p. 61.

31. Anon. (1994) TRADA data sheet: timber and brick product energy requirements. TRADA, Rickmansworth.
32. Evans, B. (1993) Counting the global costs. *Architects' Journal*, **197**(8), 57–9.
33. See reference 31.
34. See reference 21, p. 37.
35. See reference 20, p. 61.
36. See reference 32, p. 58.
37. See reference 21, p. 37.
38. See reference 32, p. 58.
39. Burek, S. and Fairbairn, A. (1993) Analysis of energy use data and building characteristics. *Proceedings of the First International Conference on Environmental Engineering, De Montfort University, Leicester.*
40. Anon. (1994) Insulation and energy. *Architecture Today*, **45**, 37–8.
41. Doggart, J. (1992) Insulation in buildings. *Energy Management*, July/August, 16–17.
42. Anon. (1990) *The Building Regulations.* L1: Conservation of fuel and power. HMSO, London.
43. Morgan, C.J. (1977) Thermal insulation, in *Specification 1976/7*, The Architectural Press, London.
44. Anon. (1994) *Warmcel Technical Data Sheet,* Excel Industries Ltd, Ebbw Vale, Gwent.
45. See reference 22, p. 55.
46. See reference 25.
47. Anon. (1991) *Rockwool Untitled Data Sheet on Fibre Size*, Rockwool Ltd, Pencoed.
48. Ridout, G. (1993) Slagging match. *Building*, Issue 40, 1 October, 46–7.
49. See reference 25, p. 50.
50. See reference 25, p. 51.
51. Maddox, S. (1993) Limiting deadly pollution. *Lighting Equipment News*, June, 12.
52. See reference 25, p. 50.
53. See reference 25.
54. Anon. (1994) CIBSE code for interior lighting. CIBSE, London.
55. See reference 10.
56. Waddington, I. (1994) Warm and dry. *Building Design*, 25 February, No. 1161, 26.
57. Anon. (1988) Energy efficiency in housing. Part 1: Code of practice for energy efficient refurbishment of housing. A designer's manual for the energy efficient refurbishment of housing. British Standards Institution, London.
58. See reference 22, p. 54.
59. Prior, J.J., Raw, G.J. and Charlesworth, J.L. (1991) *BREEAM/New homes: Version 3/91*. BRE, Garston.

Further reading

Atkinson, C.J. and Butlin, R.N. (1993) Ecolabelling of building materials and building products. BRE Information Paper 11/93. BRE, Garston.

Curwell, S. (1993) Green product specification. *Architects' Journal*, **198**(23), 30.

Evans, B. (1993) Rating environmental impact. *Architects' Journal*, **197**(20), 46.

Energy sources

CHAPTER 7

7.1 Introduction

This chapter discusses a range of energy sources, from the traditional to the newer (or rediscovered) alternatives which are beginning to be considered in building applications.

7.2 Energy

Heat and work are forms of energy which are measured in the same units and which can be compared when evaluating different energy conservation options. However, there are important differences: some forms of energy, such as electricity, are readily converted into work or into heat; others, such as fossil fuels, can be converted into heat but only part of the heat energy can provide useful work. The rest is wasted as heat losses in cooling circuits and friction.

The amount of work that can be extracted from heat energy increases as the heat source gets hotter and the cool sink gets colder. The maximum efficiency of conversion of heat into work is given by:

$$\text{Efficiency} = \frac{\text{Work extracted}}{\text{Heat available}} = 1 - \frac{T_{cold}}{T_{hot}}$$

where T_{cold} is the temperature of the cool sink (K) and T_{hot} is the temperature of heat source (K).

(As a reminder, the absolute temperature in degrees kelvin is equal to the temperature in degrees centigrade plus 273.)

For example, in electricity production, say, no more than 70% of the fuel burned at 1000 K in an engine with an air-cooled radiator at 300 K can be converted into work (1 – 300/1000 = 0.70 = 70%). In practice, combustion, friction and generation losses reduce the efficiency and electrical output further. If the temperature of combustion is increased (or the cool temperature is reduced), the efficiency increases – heat is thus more useful, and so more valuable, at higher temperatures.

Primary energy (PE) is that contained in fossil fuels in the form of coal, oil or natural gas or in nuclear energy or hydroelectricity.[1] Delivered energy (DE) is that in the fuel at its point of use after allowing for extraction (or generation) and transmission losses. The portion of the delivered energy which is of benefit after allowing for the efficiency of the consuming appliance is the useful energy (UE). Table 7.1 compares various fuels.

Table 7.1 Energy conversion efficiencies for primary to delivered and delivered to useful energy for space heating fuels

Type of fuel	*PE to DE efficiency*	*DE to UE efficiency*	*Overall (PE–UE) efficiency*
Coal	0.98[a]	0.60[b]	0.59
Gas	0.90[a]	0.70[b]	0.63
Oil	0.93[c]	0.70[b]	0.65
Electricity	0.30[a]	0.98[b]	0.29

[a]Reference 2.
[b]Estimates for a variety of devices.
[c]Reference 3.

Figures given indicate that the use of electricity, a high-grade fuel, in a low grade application such as space heating, is very wasteful of primary energy.

In theory, and generally in practice, the 'grade' of energy is reflected in the cost, as energy producers have to pass on to the consumer the costs associated with production, generation and transmission, as well as investment. However, distortions in price result from factors such as government policies and marketing; for example, the costs associated with nuclear power generation and decommissioning are still being debated, while electricity produced by nuclear power is being offered to the consumers at the same rate as that from conventional sources such as gas and coal.

Table 7.2 gives comparative costs for domestic and industrial consumers using various fuels. As can be seen industrial and other large users can use cheaper, lower grade fuels and can negotiate their tariffs and obtain substantially lower unit prices.

Fossil fuels

Fossil fuels are the partly decomposed remains of the forests and swamps that grew before and during the time of the dinosaurs. Under the high forces and

Table 7.2 Approximate comparative costs of delivered energy[4]

Application	*Coal*	*Gas*	*Heating oil*	*Electricity*	
				Peak	*Off peak*
1. Domestic heating intermittent	55[a]	28	27[b]	100[c]	48[d]
2. Industrial heating	8	10	8	55[e]	

[a]Anthracite.
[b]Kerosene.
[c]100 is the reference point for all figures in the table.
[d]The figure assumes a DE to UE efficiency of 75%.
[e]The figure is a grouped figure for a variety of tariffs.

temperatures that moved and changed the shape of the continents into their present form, these remains fossilized to become today's coal, crude oil and natural gas. Their main components are carbon and volatile hydrocarbons, but they also contain moisture, non-combustible ash and other materials such as sulphur, sodium and nitrogen. Their properties and compositions vary widely, which has led to classification systems based on factors such as carbon content and calorific value.

Gaseous petroleum fuels

These fuels are low carbon content hydrocarbons (molecules with one to four carbon atoms) which are gases at normal ambient temperatures. They are extracted from underground formations, separated from crude oils during extraction or manufactured from coal. The most common is natural gas which is used generally in the UK for heating and cooking. It consists almost entirely of methane, has very few impurities and has a high calorific value. Other gases, less commonly used in the UK, are:

- *Town gas* This is manufactured from coal and although once common, has now been superseded by cleaner, higher calorific content natural gas. Town (or manufactured) gas is approximately 30% methane, the rest being a mixture of other gases such as high carbon content hydrocarbons, carbon monoxide, carbon dioxide and hydrogen.
- *Liquid petroleum gases* (LPGs) These are hydrocarbons that are gaseous at ambient temperatures but can be liquefied at moderate pressures. Propane (three carbon atoms) liquefies at 5–15 bar and butane (four carbon atoms) liquefies at 2–6 bar. These gases are used in bottled form as an alternative to natural gas in areas outside the gas mains distribution network. LPGs are heavier than air and leaks tend to settle in lower areas, creating a possible hazard.

Liquid petroleum fuels

These fuels are hydrocarbons of medium carbon content which are liquid at ambient temperatures. The most common liquid petroleum fuels used in building applications are:

- *Paraffin* This is a light distillate heating oil used in small, free standing, flueless, vaporizing heating appliances.
- *Kerosene* This is a distillate heating oil for use in vaporizing or atomizing flued domestic heating boilers, as an alternative to gas.
- *Gas oil* This is a distillate heating oil for larger atomizing domestic and commercial burners.

There is also a whole range known as fuel oils which covers light to heavy residual or blended oils used in large industrial or commercial boilers with large preheated storage and handling equipment.

Table 7.3 Approximate properties of typical fossil fuels[5]

Item	*Net calorific value (MJ/kg)*	*Comments on impurities (%kg/kg unless otherwise noted)*
Gas		
Natural gas	39	approx. 0.001% sulphur (by volume)
Manufactured gas	19	approx. 0.001% sulphur (by volume)
Liquid petroleum gases		
Propane	46	less than 0.02% sulphur
Butane	46	less than 0.02% sulphur
Liquid petroleum fuels		
Paraffin	44	less than 0.04% sulphur
Kerosene	44	less than 0.2% sulphur
Gas oil	43	0.01% sediment, 0.01% ash, 1.0% sulphur
Coal		
Anthracite	30–35	80–90% carbon, 8% ash, 1% sulphur

Coal

Coal is classified according to its non-volatile carbon content and calorific value. In general, older coal has a higher carbon content and fewer impurities such as non-combustible ash and sulphur (and therefore has a higher calorific value) than more recent coals. Coal is also graded (and priced) according to size and the extent to which it has been washed in order to remove dust. The more common types of coal in building are:

- *Anthracite* This is premium coal with a high calorific value, lower humidity and low impurities. It is particularly useful in heating applications.
- *Smokeless fuels* These are manufactured fuels produced by heating coal (such as anthracite) in the absence of air.

Other types of coal include lignite, which is of lower calorific value, and peat, which is a much more recent product formed from partly decomposed plant residues. Table 7.3 gives data for common fossil fuels.

When fossil fuels are burned, the products released into the atmosphere include carbon dioxide (CO_2), nitrogen oxides (NOx) and sulphur dioxide (SO_2). The by-products will obviously vary with the fuel.

As we have seen in Chapters 2 and 3, CO_2 emissions are an important concern because of their contribution to global warming. Table 7.4 provides some relevant data.

The emissions due to electricity generation reflect the fuel mix and the efficiency of the plant. For example, in 1987 71% of electricity generated in the UK came from coal, 7% from oil and 4% from gas; the remainder was nuclear (16%) and hydroelectric (2%), but these do not contribute directly to CO_2 emissions. The Electricity Association anticipates reduced CO_2, NOx and SO_2 emissions as more efficient modern plant, mainly gas fired, comes on line.[7] Any significant

Table 7.4 Relationship between primary fuel use and carbon dioxide emission in the UK[6]

Fuel	*Carbon dioxide emission 1991 (kg/kWh delivered)*
Electricity	0.75
Coal	0.31
Fuel oil	0.28
Gas	0.21

change in the use of nuclear, hydroelectric and wind energy would also, of course, affect emissions. Ideally, fuel cost should take into account long-term environmental effects but at present this rarely occurs.

7.3 Combined heat and power

In main thermal power stations some 32–46% of the fuel burned is converted into electricity,[8] the remainder is dissipated as waste heat in chimneys, cooling circuits and in the generator itself. Power stations are normally sited far from densely populated areas and so the waste heat is lost. Further losses are incurred during transmission to the points of use.

In combined heat and power (CHP) or co-generation installations, gas or diesel internal combustion engines (or, in large installations, gas turbines with reheat) drive a generator and produce electricity. Typically 23–28% of the fuel used can be converted into electricity in this way. Of the other 72–77%, some 63–66% is available as waste heat suitable for normal space and hot water service heating so that most of the energy in the fuel is useful, as shown in Figure 7.1.

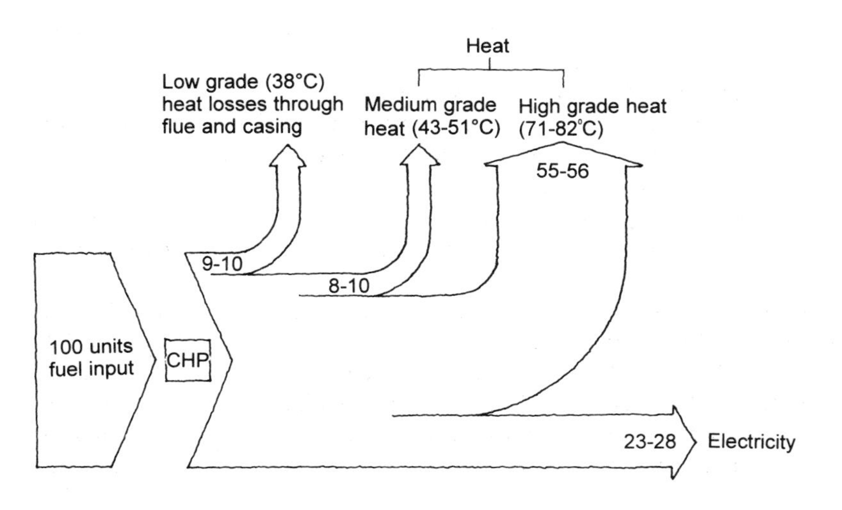

7.1 Typical energy balance of a CHP unit.[9]

(a)

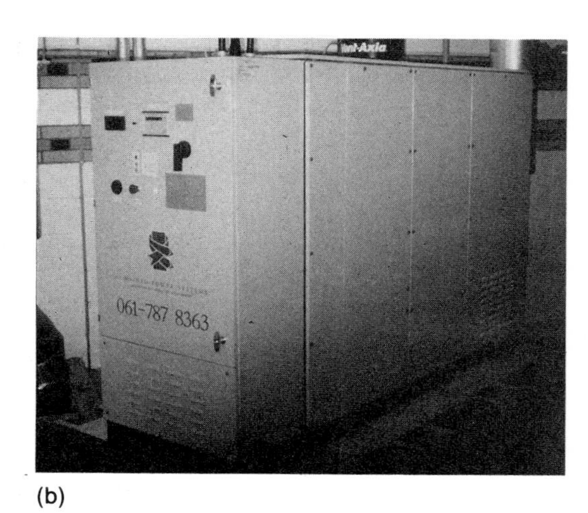

(b)

(a) Typical unit.
(b) Installation at De Montfort.

7.2 CHP installations.

These systems were developed from traditional standby generating sets, but there is now considerable expertise in gas-fired CHP units with an electrical output of 15–300 kW running at the same time (i.e. in parallel) with the normal electrical supply. The mains supply meets any electrical demand in excess of the capacity of the CHP unit and also absorbs any spare capacity. The heat from the cooling circuits is used for space and hot water service heating and is treated similarly to a conventional boiler. Figure 7.2 shows a typical unit with the panel removed and on the right the actual installation with an acoustic enclosure at the De Montfort University Queens Building (Chapter 13). Other options such as multiple (and/or larger) units, emergency operation, and other fuels are also possible.

Some points to be considered when planning an installation are:

- High overall efficiencies can only be achieved if both heat and electricity are used. The unit(s) must be run for long periods to be cost-effective (say 3500–4500 hours per year), thus, the unit must be relatively small or excess heat will need to be rejected. Consequently, CHPs are most suitable for providing a base load.
- It is now possible to sell surplus electricity to the utilities or to rent the grid to transport surplus from one site to another. Tariffs and charges are such, however, that the rate for displaced 'imports' of electricity are always significantly greater than the rate for 'exports': the economics of this bias favour smaller CHPs where all output is used on site.
- Maintenance and service costs are significant and must be fully considered, allowing for regular replacement of components and major servicing over the lifetime of the installation.

7.4 Heat pumps

A heat pump is a device that transfers heat from a cold source to a hot one. As this is opposite to the normal direction of flow, energy is required to transfer the heat. The energy used in this process is dissipated as heat, and adds to the heat extracted on the cold side. The equation for the heat balance is:

$$Q_{hot} = Q_{cold} + W$$

where Q_{hot} is the heat rejected at the hot source, Q_{cold} is the heat extracted from the cold source and W is the energy used in transfer.

A domestic refrigerator is similar in that heat is removed from the inside of the refrigerator, thus cooling it, and is rejected to the room which is warmer. In a heat pump the emphasis, however, is normally on obtaining the maximum amount of heat for the warm room.

The coefficient of performance (COP) for a heat pump is given by:

$$COP = \frac{Q_{hot}}{W}$$

with Q_{hot} and W defined as above. The COP depends on the temperature of the cold and heat sources – the maximum COP is given by:

$$COP = \frac{T_H}{T_H - T_L}$$

where T_H is the absolute temperature of the high temperature heat source (K) and T_L is the absolute temperature of the low temperature heat source (K).

In building applications heat is usually extracted from ambient air. The use of ground or surface water (where available) improves the performance as in mid-winter water is usually warmer than the ambient air.

In building applications the COP is typically 2.5–3.5; that is to say, for each unit of energy used to drive the heat pump, 2.5–3.5 units of useful heat are delivered.

Heat pumps in buildings are usually powered electrically, and some of the energy savings will be lost during electrical generation and transmission. They can also be driven by oil- or gas-fired engines, but this is more applicable to larger installations.

While this may seem attractive, there are further considerations, which mean that the environmental 'acceptability' of heat pumps is not evident. Firstly, for those that run on electricity the greater CO_2 production that may result in comparison with, say, a gas-fired condensing boiler (Table 7.4) is a disadvantage; the higher monetary cost of electricity may also mean that heat pumps are not economic. Another factor is that most current heat pump designs use CFCs or HCFCs and thus there is a further environmental disadvantage. Alternative refrigerants for heat pumps could be developed, as could other heat pump systems. There is potential, but progress is slow.

7.5 Renewable sources of energy

Approximately 70% of current UK energy consumption is derived from solar energy in the form of fossil fuels, formed by photosynthesis, accumulated, concentrated and stored over millions of years. The size of the reserves may be

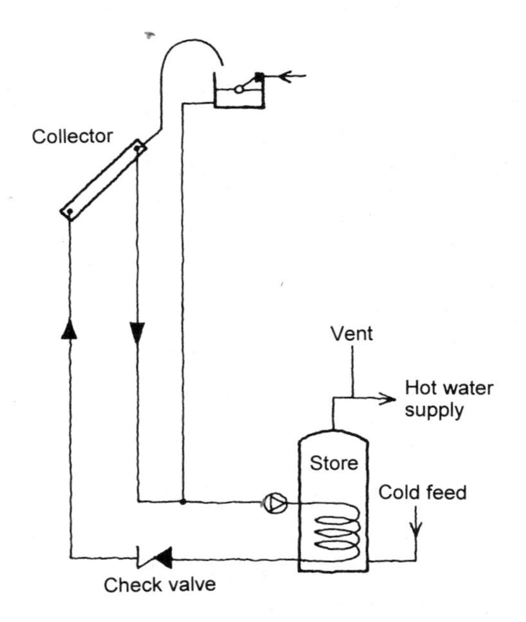

7.3 Simplified schematic of a typical solar water heater system (indirect, vented, pumped circulation).[11]

debatable, but this energy use represents a constant depletion of resources and at some future time this source will be exhausted.

Renewable sources, all ultimately derived from the Sun's energy, include solar power (both active, i.e. incorporating moving water or air, and passive systems and photovoltaic devices), wind and wave power and biological sources such as wood and fuels derived from crops. They are renewable (or, similarly, sustainable) because the Sun will continue to provide their energy.

However, the energy density of solar and wind energy is not great and the availability is very variable. To overcome these disadvantages, more research and development are required. In the UK some impetus has come from the obligation on regional electricity companies to generate some of their electricity from non-fossil fuels. The government has set a target of a contribution of 1500 MW of renewable electricity capacity by the year 2000[10] (or about 3% of the current maximum electrical load).

Active solar heating

Active solar heating uses collectors to convert solar radiation into heat for space and hot water heating. Collectors are surfaces painted black to absorb most of the incoming radiation, glazed and insulated to reduce the heat losses and suitably orientated (usually within 30° or so of due south) to optimize the amount of energy incident on them.

The heat generated in the collectors is removed and circulated by a water circuit (or less commonly air). A typical installation usually also includes some form of heat storage such as a hot water store, as illustrated in Figure 7.3. There is also likely to be a means of providing back-up heating when solar heating is insufficient.

7.4 Solar water heating at a Florida alligator farm.[14]

There are many types of collectors for different applications. Where heat at low temperatures is adequate, such as for swimming pools, the collectors can be simple, unglazed, uninsulated, and therefore relatively cheap. Where the heat is needed at higher temperatures, such as for space heating, or where collectors are required to operate at lower ambient temperatures, the heat losses become greater and the collectors must be insulated and glazed. In more advanced collectors the heat losses are further reduced by special surface treatments known as selective coatings that reduce reradiation from the collector while retaining a very good absorption of solar radiation. Even lower heat losses and higher temperatures can be achieved by enclosing the collectors in a vacuum (evacuated tubular collectors) and, beyond that, by focusing incident solar radiation into a smaller area, as in concentrating collectors.

Some 2–13% of the monthly solar radiation incident on a collector can be delivered as useful heat in solar water heating applications in the UK. This amounts to some 670–1000 MJ/m^2 per year;[12] the precise figure depends on factors such as water and ambient air temperatures. At current prices the capital cost of such an installation for domestic hot water heating is generally not cost-effective in the middle and northern European (economic) climate. Space heating and higher temperature applications are even less viable. Swimming pool heating and solar water heating are, however, becoming established in sunnier climates and where traditional fuels are unavailable or expensive.

Figure 7.4 shows a system at an alligator farm in Florida, where solar-heated water (at a temperature of 60 °C) is blended with well water at 22 °C which is required for washing down the buildings that house the alligators.[13] By maintaining a constant 32 °C for the buildings and all water used, the alligators experience no thermal shock and grow very well.

Photovoltaics

When light falls on certain materials known as semiconductors, the energy of the incident light (photons) frees some of the electrons in the materials. A difference in potential develops within the material thus providing a direct conversion of energy into electricity. These photoelectric cells are typically manufactured from silicon, grown either in single crystals which improves the performance at increased costs or, more commonly, as a polycrystalline or amorphous silicon deposit into a substrate, which gives lower efficiencies but much lower costs. The photoelectric cells are then interconnected in groups with diode protection for undervoltage and encapsulated into panels to provide protection from the environment.

Solar radiation is subject to the vagaries of the weather, as well as daily cycles, and so most installations must include batteries, a charge controller and, if the output required is AC, an inverter.

The conversion efficiency of a photoelectric cell is typically 12–15%, but losses within the control system and degradation in performance can reduce this to approximately 9–12%. Depending on the method of construction, the cells themselves may cover only 70% of the area of the panel. Assuming a peak incident solar radiation of, say, 600 W/m^2, we have approximately 45 W/m^2 peak from the cells.

It has been estimated[15] that a photovoltaic system with 66 m^2 will generate 2600–3300 kWh/yr. By comparison, an average three-bedroomed dwelling might consume about 2000 kWh/yr (Table 3.3).

Photovoltaic energy is now well established in remote areas without access to the grid for high value applications such as beacons, communications equipment, medical refrigeration and special lighting, but is generally too expensive for other applications. However, developments in semiconductor manufacturing techniques and growing markets in other areas, such as providing power for calculators and boat lights, are reducing costs and at present there is great interest in planning for buildings with areas of photovoltaic wall.

Wind power

Air motion requires energy: the Sun and the rotation of the Earth provide the energy to move large air masses, thus producing wind, and this 'wind' energy can be harnessed by sails to move a boat or to turn a windmill to pump water or grind flour, or, more recently, generate electricity.

The power that an ideal windmill can extract from the wind is given by the following formula, derived by Betz:[16]

$$P = 0.645(A \times u^3)$$

where P is the power (W), A is the swept area of the wind turbine (m^2), and u is

Table 7.5 Comparison of power extractable from the wind for windmills in London and at an exposed site[17,a]

	Wind speed[b] (m/s)	*Terrain correction factor*	*Resulting speed (m/s)*	*Power extractable (W/m²)*
Urban house	4.0	0.62	2.48	9.84
Exposed house on coast	5.5	1.0	5.5	107.3

[a]Assumed height of 10 m.
[b]This is the mean wind speed exceeded 50% of the time.

the wind speed (m/s). Aerodynamic, mechanical and electrical losses are likely to reduce this by one-third.

Wind speed varies constantly in magnitude and direction. This is normally presented as a frequency distribution curve, which indicates the percentage of the time that a given wind speed is exceeded. Since economics dictates that equipment must operate during most of the year to repay the investment, most windmills are designed to operate at common wind speeds of 5–15 m/s (Appendix A). Batteries (or the grid) provide power when speed is low and safety devices protect the mill in high winds.

If we take a wind speed of 5 m/s we see that the power an ideal mill can extract is about 80 W/m^2. This is relatively low, but roughly of the same order of magnitude as solar energy collectors.

Winds are also slowed by friction with the ground and deflected by the topography and surrounding buildings, therefore the mean wind speed data must be corrected for these effects.

Table 7.5 compares the power an ideal windmill can extract at the site of an urban house in London and at the site of an exposed house near the coast in Cornwall.

In this example, the power an ideal windmill can extract in the coastal area is therefore about 10 times greater than in the urban area, thus showing the importance of wind speed.

At present, wind power generation applications are mainly limited to:

- remote locations without access to the grid (i.e. where alternative sources of electricity are equally costly and/or restricted), particularly for small essential loads (e.g. communications equipment), or where the vagaries of the wind are not critical (e.g. base load lighting), and
- large installations in exposed locations where economies of scale can be applied (i.e. wind farms, as shown in Figure 7.5).

In the UK, onshore wind generation from medium- to large-turbine (400 kW) wind farms is considered one of the more promising renewable energy sources, with the potential to provide some 30% of current electrical energy demand.[19] The government plan is to generate 10% of the demand by the year 2025, which would require some 4000 km^2 of land (agricultural use can be retained), although

7.5 Wind farm.[18]

opposition from pressure groups may reduce this target figure. The current estimated generation costs are similar to domestic tariff rates but a higher price is often set as an incentive to investors.

Hydroelectric power

Hydroelectric power relies on the energy released when water moves from high to low levels, much as wind power relies on air movement from regions of high to low pressure. However, the energy density of hydroelectric power is greater than that of the wind because of the higher density of water (1000 times greater than air), the higher pressures available, and the concentrating effect of water in rivers and lakes. Consequently, hydroelectric power generation has been a conventional source of energy for some time; water power for milling has, of course, been used for much longer.

In the UK, hydroelectric generation accounts for under 2% of the total electricity supply; most comes from large-scale sites in Scotland. The potential of the unexploited resources is limited by economic and environmental constraints, but it is estimated that the generating capacity of sites in excess of 5 kW could be increased by 50%.[20]

With developments in low-cost controls and the relative increase in the cost of energy, mini- and micro-hydroelectric generation is becoming increasingly viable, and it is now easier to exploit previously uneconomic water courses with smaller flows or falls. In the short to medium term, however, micro-hydroelectric generation in the UK will remain an option only for private schemes for small communities isolated from the grid and near to water courses. The technology is well established, however, and the potential for other parts of the world is promising.

Table 7.6 Typical characteristics of dry agricultural waste fuel[15]

Item	*Net calorific value (MJ/kg)*	*Comments on impurities*
Harvested straw	15[a]	15% moisture content
Timber harvest waste	10[b]	55% moisture content; ash 1–2%kg/kg; low sulphur content

[a]At 15% moisture content.
[b]At 55% moisture content.

Wood and straw

Wood was once the traditional fuel for heating and cooking, and is still widely used in many parts of the world. In many countries it was replaced when more efficient, controllable, convenient and less locally polluting forms of heating became available. Modern wood-burning stoves, however, can be efficient and control has improved. Automatic fuel loading and removal of ashes are usually restricted to large installations.

Agricultural waste, such as straw bales from cereal crops and forestry waste such as branches and tree tops from thinning and timber harvesting, can also be used as fuels. Because the calorific value is low they are an economic alternative only where they are locally available at low cost; sometimes they are used in dual-fuel installations with, say, oil. Typical data is given in Table 7.6.

As we have seen, in photosynthesis green plants use sunlight to convert carbon dioxide in the atmosphere (or dissolved in water) into oxygen and fixed carbon. When the carbon compounds are burned, the chemical energy in the compounds is released and the carbon is emitted as carbon dioxide, so that over the lifetime of the plant there is no net increase in carbon dioxide. For energy crops, selected and planted to provide fuel, the annual energy yield available is estimated at 130–700 GJ/ha.[22] Worldwide, an estimated 5.5 million hectares are expected to be surplus to requirements for food production by the year 2010,[23] and some of this area could be used for energy crops.

Other

Other sources, including geothermal energy, biogas and municipal waste, may have a greater role to play in the future.

Guidelines

1. The use of high-grade fuels such as electricity in low-grade applications – for example, heating – is wasteful of primary energy.
2. Fuels have impurities that become pollutants, so fuels must be chosen carefully.
3. Combined heat and power (CHP) can be very energy efficient but must be judged on its merits for each application.

4. Solar energy is not presently economic for most applications in middle and northern Europe.
5. There is growing interest in photovoltaics as the technology develops and prices fall.
6. Wind power is now technically proven, but environmental concerns will be a major factor in its development.

References

1. Bakke, P. *et al.* (1975) Energy conservation: a study of energy consumption in buildings and possible means of saving energy in housing. BRE Current Paper 56/75. BRE, Garston.
2. Shorrock, L.D. and Henderson, G. (1990) Energy use in buildings and carbon dioxide emissions. BRE, Garston.
3. See reference 1.
4. Based on private conversations between R. Godoy and fuel merchants and supply companies.
5. Anon. (1988) *CIBSE Guide C5: Fuels and Combustion*, CIBSE, London.
6. Prior J.J. *et al.* (1991) *BREEAM/New homes. Version 3/92.* An environmental assessment for new homes. BRE, Garston.
7. Anon. (Undated – approximately 1993) *The Benefits of Electricity – An Overview*, Electricity Association.
8. Ibid., p. 14.
9. Forrest, R., Heap, C. and Doggart, J. (1985) Small-scale combined heat and power. *Energy Technology*, Series 4. Energy Efficiency Office, Harwell.
10. Anon. (1993) *Renewable Energy*, Bulletin No. 5. Information on the non-fossil fuel obligation for generators of electricity from renewable energy sources. Department of Trade and Industry, London.
11. Wozniak, S. (1979) *Solar Heating Systems for the UK: Design, Installation and Economic Aspects,* Department of the Environment Building Research Establishment. HMSO, London.
12. Anon. (1989) Solar heating systems for domestic hot water. BS 5918: 1989. British Standards Institution, London.
13. Healey, H.M. (1994) A modular solar system provides hot water for alligator farm. *ASHRAE Journal*, **36**(3), 53–5.
14. Ibid.
15. Anon. (1994) *An Assessment of Renewable Energy for the UK*, Energy Technology Supply Unit. HMSO, London.
16. Golding, E.W. (1955) *The Generation of Electricity by Wind Power*, E & FN Spon, London.
17. Anon. (1988) *CIBSE Guide A2: Weather and Solar Data*, CIBSE, London.
18. Anon. (1994) *British Wind Energy Association*, BWEA, London.
19. See reference 15.
20. See reference 15.
21. See reference 15.
22. See reference 15.
23. See reference 15.

Further reading

ETSU (1985) *Energy Technology Series* No. 5: Heat pumps for heating in buildings. Energy Efficiency Office.

Lighting

CHAPTER 8

8.1 Introduction

As noted in Chapter 3, artificial lighting in an office building accounts for approximately 16% of its energy consumption. For other building types the figures will vary but it is almost always significant.

Substituting increased natural daylighting (daylight includes sunlight, which is the direct beam, and skylight, i.e. visible diffuse sky radiation) for artificial lighting in offices and other buildings offers large potential energy savings. This is true provided overheating and glare can be controlled and provided significant increased heat losses do not result.

8.2 Daylighting

Most people prefer daylight. The contact with changing natural light is physiologically, psychologically and architecturally important. Le Corbusier said 'architecture is the masterly, correct and magnificent play of masses brought together in light...'[1]

Daylight availability varies enormously (in this way it is very similar to natural ventilation) and is a key design issue. (Temperature variations, on the other hand, are more seasonal and are therefore easier to control; noise level variability depends very much on the site). The average levels of daylight are given in Appendix A.

The lighting level in the space is very important. One aspect of this is the daylight factor, which is defined as the illuminance received at a point, indoors, from a sky of known or assumed luminance distribution, expressed as a percentage of the horizontal illuminance outdoors from an unobstructed hemisphere of the same sky; direct sunlight is excluded from both values of illuminance.[2]

Recommended daylight factors, as shown in Table 8.1, exist but need to be treated with caution, because they are not in fact high enough if the optimum use of daylight is to be made. The best guidance is probably to say that daylight should be maximized subject to the constraints of glare, increased solar gains and possible greater heat loss. (We shall return to this issue below.) There are also recommended lighting levels for spaces as shown in Table 8.2.

It should be noted that the values given in Table 8.2 are guidelines and judgement should be used. For offices, in particular, there is the possibility of lowering the level to, say, 300 lux, provided task lighting can supply 500 lux (or more) where required.

Table 8.1 Recommended daylight factors[3]

Space	Minimum (%)	Average (%)
Lounges in dwellings	0.5	1.5
School classrooms	2	5
Offices: general	2	5
Hospital wards	1	5

The amount of light that enters a space obviously depends on the areas and disposition of the glazing. To a large extent the amount of daylight at a point in a room depends on the area of sky that can be seen through the window. Thus, there tend to be wide disparities in natural light levels between areas close to windows and those some distance from them. In Appendix C a very simple calculation for the average daylight factor is given. Numerous more sophisticated calculation procedures exist and computer simulations are beginning to produce very accurate images of internal light conditions.[5]

It is common to apply the daylight factor to what is known as the standard overcast sky illuminance of 5000 lux.[6] This value is exceeded about 85% of the standard working year (Table C.2). Thus, under a standard overcast sky a daylight factor of 10% at a point near a glazed wall would give an illuminance of 500 lux. This indicates that if you had a desk at that position in a general office the light level would normally be sufficient, but for 15% of the working hours

Table 8.2 Recommended lighting levels[4]

Space	Standard maintained illuminance (lux)
Atria	
– general movement	50–200
– plant growth	500–3000
Assembly shops	
– medium work	500
Lecture theatres	300
Newsagents' shops	500
Offices	
– filing rooms	300
– general offices	500
Paint works	
– colour matching	1000
Public rooms, village halls, church halls	300
Teaching spaces	300

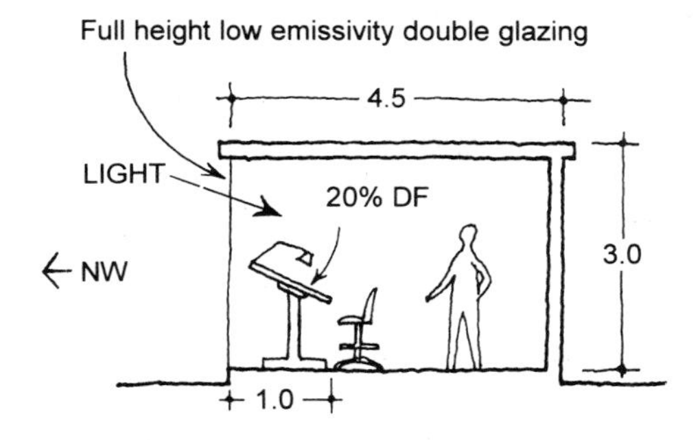

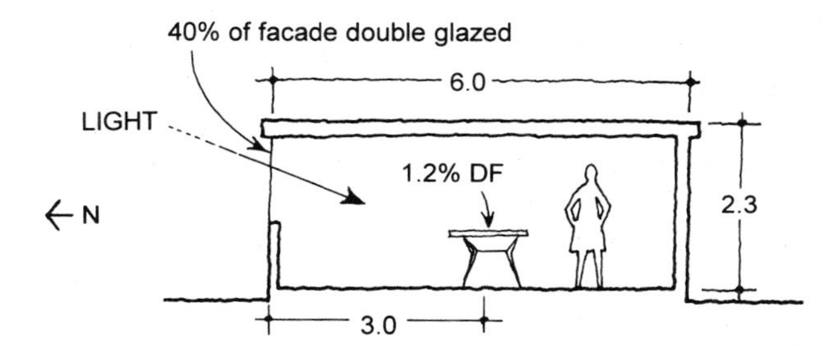

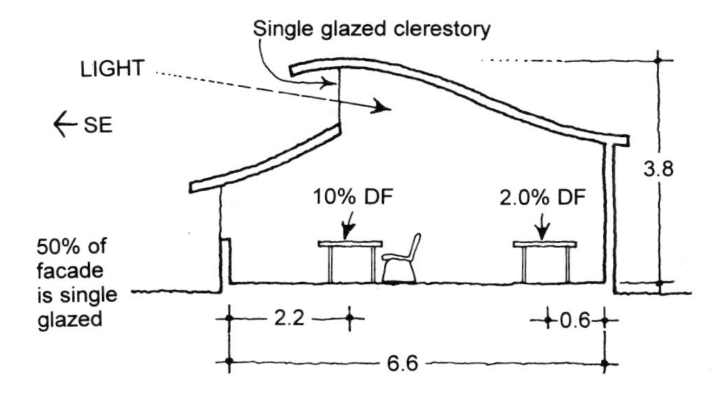

8.1 Indicative daylight factors.

you would need to turn some lights on to maintain the recommended lighting level. It also shows why the 'recommended' daylight factors of Table 8.1 are too low. Could the lighting at times be too bright? In one way the answer is 'no' since most people will happily read outside on the sunniest summer day (the danger here being excessive exposure to UV radiation), but there is the possibility of glare, and the very real problem of potential thermal discomfort due to direct solar gain at such a position demands some form of solar control.

Although the main reason to make effective use of daylight is to reduce artificial energy consumption there are also potentially useful heat gains available. A rule of thumb is that an illumination level of about 1000 lux outside would correspond to total solar radiation of about 10 W/m^2 on a horizontal surface. (This is based approximately on the highest daylight level of approximately 100 000 lux, corresponding to about 900 W/m^2 on a horizontal surface.)

Making effective use of daylight depends very much on planning the building. As discussed in Chapter 2, highly articulated spaces with a greater perimeter length will normally offer more potential daylight (and natural ventilation) but at the possible cost of greater heat loss. The art lies in finding the right balance – one key element of which is reducing the heat loss at night (see, for example, Figure 8.3).

The more glazing at the perimeter wall, the higher the daylight factor. Figure 8.1 gives a more quantitative idea of measured daylight factors in three cases.

A key point for all these examples, and for many of the more innovative daylighting systems, is that ceiling reflectances must be kept high by using very light colours (Table 8.3); another consideration is that the ceiling should not be encumbered by bulky artificial lighting systems. High ceilings with windows running up to them, as in Victorian schools and hospitals, help to increase light levels (and produce good uniformity – see below). For very rough calculations, Figure 8.2 gives some rule of thumb guidelines.

Obviously, wherever possible, activities that need a great deal of light should be placed near the perimeter. Traditionally, this has been done in weavers' cottages and large buildings. (In the author's offices – a nineteenth-century former piano factory – glazing occupies 35% of the perimeter wall area and pianos were polished close to the windows in an area with a daylight factor of about 6%.) Similarly, a variety of spaces that need lighting only occasionally (such as storerooms), or that have lower lighting requirements (such as circulation spaces) should be moved towards the interior.

Table 8.3 Approximate values of light reflectance[7]

Material	*Reflectance*
1. *Internal*	
White paint[a]	0.85
White paper	0.8
Light grey paint	0.68
Strong yellow paint	0.64
Wood – light veneer	0.4
Strong green paint	0.22
Quarry tiles	0.1
Carpet – deep colours	0.1
2. *External*	
Snow (new)	0.8
Portland stone	0.6
Sand	0.3
Brickwork (red)	0.2
Green vegetation	0.1

[a]BS 4800 colour codes are given in the original reference.

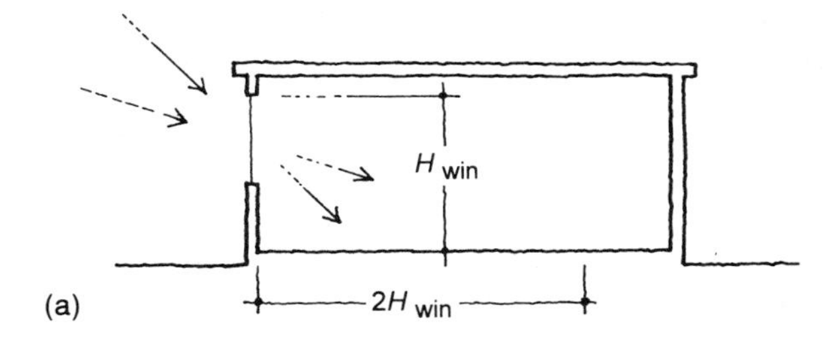

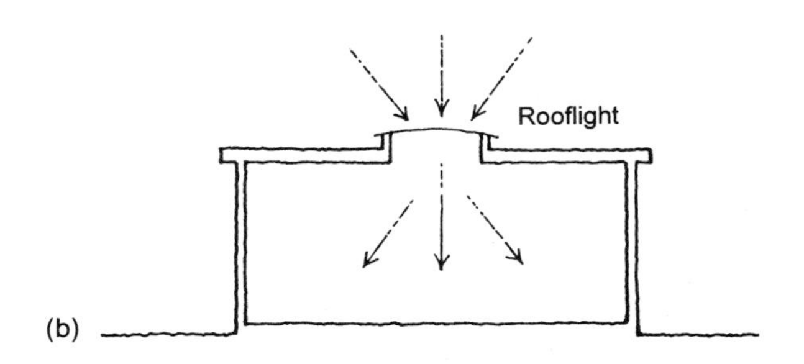

(a) Sidelighting: average DF $= 20(A_g/A_f)\%$ (in the area adjacent to the window to about $2H_{win}$ away).
(b) Horizontal skylight: average DF $= 50(A_g/A_f)\%$, where DF is the daylight factor as a percentage, A_g is the area of glazing (m^2), A_f is the area of floor to be lighted (m^2), and H_{win} is the window head height (m).

8.2 Very approximate average daylight factors for single glazing.[8]

One standard reference suggests that when the average daylight factor exceeds 5% on the horizontal plane an interior will look cheerfully bright, and when the factor is below 2% the interior will not be perceived as having adequate daylight and electric lighting may be in constant use.[9] Achieving 5% (and, often, ideally more) is by no means easy in many situations and requires careful design.

The quality of lighting and the perceived need to switch on artificial lighting depends much on the range of light levels from the front to back of the space. Too great a range can give a predominantly gloomy character to the space. One criterion[10] for acceptable uniformity in spaces with windows on one wall is that $(d/w + d/h)$ shall not exceed $2/(1 - R_b)$ where d is the depth of the room, w is the width of the room, h is the height of the window head above floor level and R_b is the area weighted average reflectance of the half of the interior remote from the window. For single-storey buildings, or the uppermost floor of multi-storey buildings, rooflights can provide additional lighting away from the perimeter. Figure 8.3 shows an example at a Hampshire school, in which the rooflights were fitted with movable insulated shutters to reduce heat loss at night.

Rooflights increase uniformity and also have the advantage of 'seeing' more of the sky than vertical glazing. The sky is also brighter directly above us than it is at the horizon, and this is a further advantage for rooflights.

On a standard overcast day the approximate lighting levels just outside a vertical window and outside a horizontal skylight are about 2000 lux and 5000 lux, respectively. Thus, it makes sense to consider the use of skylights where possible, always keeping in mind that solar gain needs to be kept under control. Thought also needs to be given to the effects of toplighting because some people find that this gives objects a duller appearance owing to a lack of modelling (see section 8.3 below for a further discussion). Another consideration is that the heat loss of a

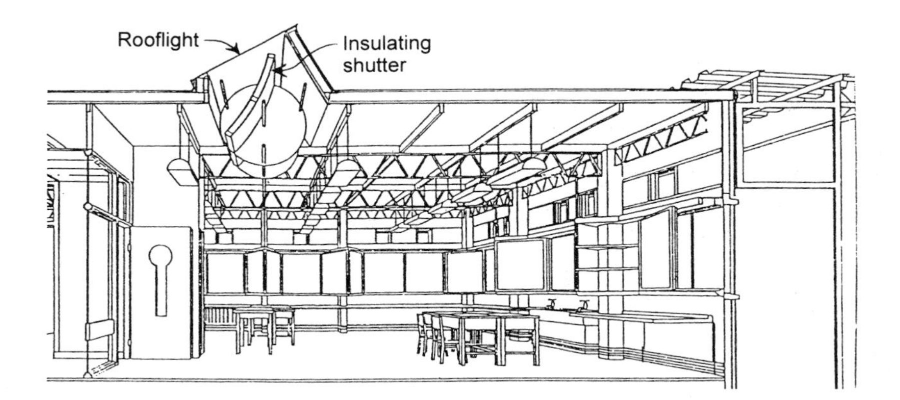

8.3 Rooflights at Crookham Church School. (Architects: Edward Cullinan Architects.)

rooflight on a clear night can be greater than that of a window because the rooflight 'sees' more of the cold sky than the window and so radiation loss is greater; losses due to conduction/convection can also be somewhat greater with rooflights.

In more common applications where only sidelighting is possible, one way of attempting to improve uniformity has been the use of lightshelves (Figure 8.4) and other reflective mechanisms such as prismatic glass; a number of innovative daylighting systems are reviewed in Reference 11.

The BRE has carried out extensive tests on lightshelves in south-facing walls and found that in sunny conditions they had the advantage of shading an area of the room close to the windows from direct sunlight but resulted in some light loss and provided relatively small redirection of light.[12] In overcast skies light levels are reduced by 5–30%, depending on the position in the room with the 5% loss being at the back of the room. This pattern of some light loss and some redistribution in sunny conditions and light loss in overcast conditions was also found to be valid in the same study for similar techniques such as mirrored louvres and prismatic films.

One point to keep in mind when evaluating such systems is the relative pro-

8.4 Typical lightshelf.

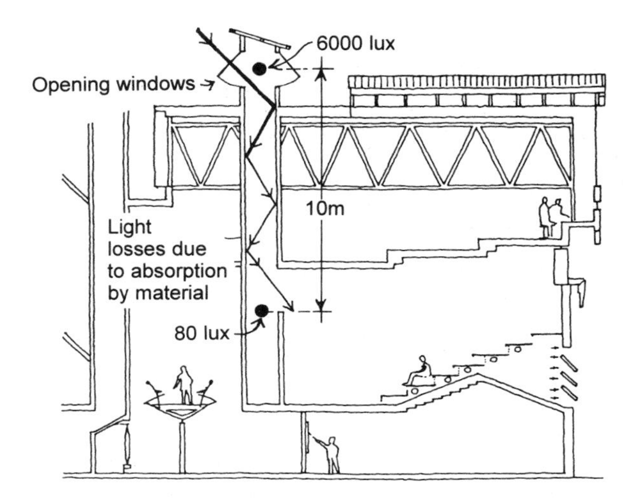

8.5 Representative measured light levels in tall shafts at De Montfort University's Queens Building.

portion of times when it is overcast and when it is sunny. In the UK it is often overcast – the number of hours of bright sunshine in London averaged over the year is only about four per day and in the winter when light levels are lower the figure is less than half this (Appendix A).

There may, however, be other ways of looking at lightshelves that make them more promising. One possibility is to use them as a perimeter zone for running services. Another is to have an adjustable lightshelf similar to a Venetian blind. In sunny conditions the blind would be closed to reflect light up to the ceiling; under overcast skies it would be open allowing more light to the zone near the window (and eliminating the losses that would otherwise occur at the ceiling).

If we now look at attempting to provide daylight or sunlight from rooftop level to floors lower down we find it is not easy. In atria with large glazed areas and few obstructions it is not a problem but shafts are much less effective. Measurements of the internal light levels at the top and bottom of the stacks at De Montfort (Chapter 13) show losses of 98–99% as the light travels down the stack. Figure 8.5 shows this schematically.

The De Montfort shafts are diamond-shaped (3.2 m between the most distant vertices and 1.4 m between the other two) and lined with a dull white fabric (which covers the acoustic attenuation). They were not specifically designed as light shafts and by altering the geometry and using smoother and whiter surfaces reflective losses could be reduced and light levels at the base could be improved somewhat but not greatly.

Larger shafts have the disadvantage of requiring more space for openings on intermediate floors. A common problem with all shafts is that with time the surfaces become less clean and less reflective. Additionally, the light that eventually reaches the space may have a curiously dead quality even on sunny days because the direct sunlight is effectively transformed into diffuse light by reflections.

More highly technological systems of light pipes or piped sunlighting with slow-tracking lenses or mirrors exist[13] and others are being developed. However, it must be noted that in many areas of the world, including much of northern Europe, systems designed particularly for direct sunlight are not likely to be appropriate as sunlight is often in short supply. Furthermore, the systems cannot enhance the diffuse light and, indeed, reduce it. It would seem, therefore, that these solutions with their higher costs are destined for a minority of buildings in sunnier climates.

8.3 Artificial lighting

Daylighting and artificial lighting

The artificial lighting scheme and daylighting should complement each other. One important aspect of this is the quality or spectral composition of the light, which varies with a number of factors including the position of the Sun and weather conditions, the most obvious being cloud type and cover.

In order to allow discussion and comparison of light sources, the concept of a correlated colour temperature (CCT) has been developed. The CCT can be considered, somewhat simplistically, to relate the temperature of a radiating body or object and the colour of the light it produces. (Fuller discussions can be found in Reference 14.) By international agreement the CCT of average daylight (sunlight plus skylight) outdoors is taken as 6500 K (in practice it varies between 4000 and 12 000 K).[15]

The CCT allows light sources to be located on a simple warm–cold scale. If the CCT is less than 3300 K the source is considered warm; if it is between 3300 and 5300 K it is intermediate; and if it is above 5300 K it is cold. If this seems paradoxical, consider the warm glow from a dying open fire compared to the 'white' heat from much hotter temperature sources such as molten metal.

It is, of course, very difficult to obtain quantitative levels of natural daylighting with artificial sources (although in hospital operating theatres local light levels can reach 50 000 lux) but for our normal activities this is fortunately not necessary. Another question, however, is: How close can we come to reproducing the quality of daylight? In Figure 2.4 we saw a typical distribution spectrum for solar radiation at the Earth's surface but, of course, solar radiation and daylight vary continuously. The CIE (Commission Internationale de l'Eclairage, or, in English, the International Commission on Illumination) has defined a set of reference illuminants mainly based on a series of spectral power distributions of phases of daylight.[16] The CIE colour rendering index (CRI) is a measure of how accurately colours of surfaces illuminated by a given light source match those of the same surfaces under one of the reference illuminants. The closer the CRI is to 100, the better the agreement.

Figure 8.6 gives spectral power distribution charts (i.e. the amount of light generated in each band) and CRIs for two types of fluorescent lamp. The shapes of these curves can be compared with that of north sky daylight, as shown in Figure 8.7 (for interest, the distribution is also shown in Figure 2.4).

Although individual lamps have their own CRIs it is common to consider groups. CIE colour rendering group 1A (CRI of 90 or more) is commonly used

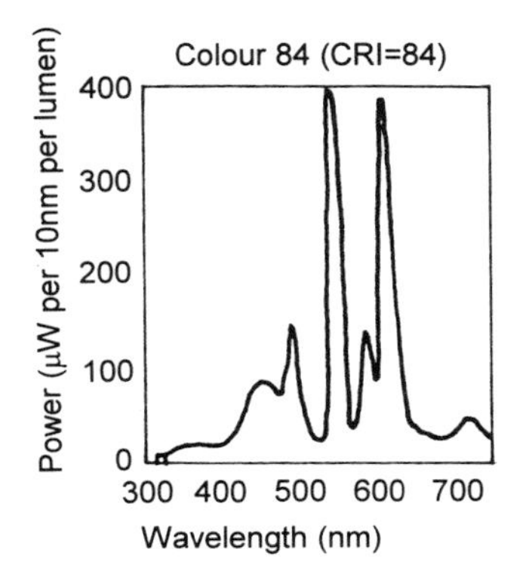

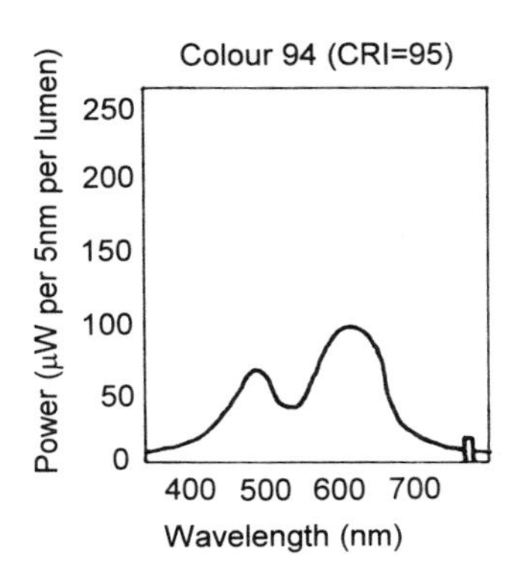

8.6 Spectral power distribution charts.[17]

where accurate colour matching is required, group 1B (CRI from 80 to 90) where good colour is needed for reasons of appearance and group 2 (CRI from 60 to 80) where moderate colour rendering is required. Groups 3 and 4 give worse colour performance.

What can be done to ensure that daylight and artificial light complement each other, or are at least mixed successfully, always remembering, of course, that the spectral composition of the daylight reaching indoors will be a function of the glazing, the colours of the interior, the shape of the room and so forth? As a simple experiment, switch on a tungsten table lamp (CCT of 2700 K) on a white tablecloth in broad daylight and note the yellow glow that contrasts with the cool daylight. A number of architects favour lighting that is warm but not too yellowish. The CIBSE way of expressing this is their recommendation that, in general, room lighting discrepancies between the colour of electric light and daylight can be reduced by using lamps of intermediate colour temperature (3300–5300 K).[19] Creating transition zones from mainly naturally lighted areas to more artificially lighted ones is part of the art of architecture.

A second issue in artificial light complementing daylight is how light can model objects, i.e. show their texture and form. A completely uniform distribution of

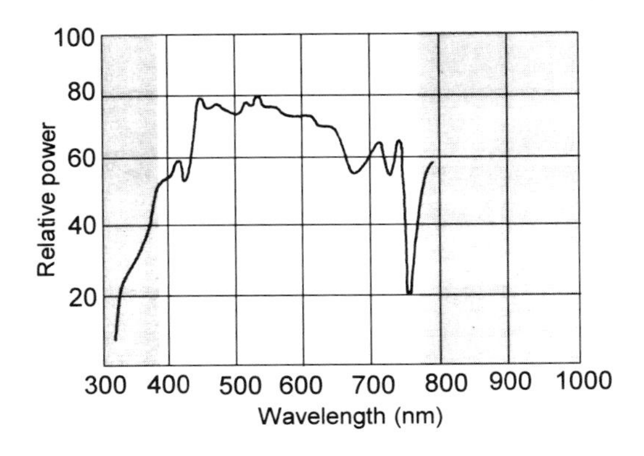

8.7 Spectral composition of north sky daylight of 5700 K (*NB*: infrared not shown).[18]

Table 8.4 Light source data[a]

Light source	*Approximate (lm/W)*	*Approximate colour temperature (K)*	*Colour rendering group*	*Approximate lamp life (h)*	*Capable of being dimmed*	*Starting*	*Comments*
1. General lighting service GLS (filament)	10–20	2700	1A	1000	Yes	Prompt	Energy inefficient; produces a great deal of heat
2. Tungsten halogen (TH) (filament)	10–20	3000	1A	2000	Yes	Prompt	Low voltage TH tends to be used for shop displays or commercial developments for effect.
3. Tubular fluorescent (MCF)	60–100[f]	3 categories a. Less than 3300 b. 3300–5300 c. Greater than 5300	Variable[d]	6000–12 000 (with high frequency ballasts 7500–15 000)	Yes	Prompt	Controls incorporating high frequency ballasts improve the energy efficiency of fluorescent lamps slightly but are perhaps more important as an increased source of comfort since they reduce flicker.
4. Compact fluorescent[b]	50–80[f]	2700–6500	1B	8000–10 000	Yes	Sluggish	Starting performance is improving
5. Metal halide[b]	70–80	3000–6000	1A–2	5600–13 000	Limited dimming available	1–2 minute run-up	
6. High pressure mercury (MB[b])	40–60	3300–4000	3	14 000–28 000[c]	No	2–5 minute run-up	
7. High pressure sodium (SON[b])	60–120	2000–3000	1B-4	14 000–28 000	Limited dimming available	1.5–6 minute run-up	
8. Low pressure sodium (SOx)	100–190		[e]	11 500–23 000	No	8–12 minute run-up	

[a]Compiled principally from Reference 23.
More complete and detailed information may be found in these references.
[b]Numerous designations exist. See Reference 23.
[c]Blended high pressure mercury lamps have shorter lamp lives.
[d]See Reference 23, p. 96 and manufacturers' literature.
Colour rendering groups 1A, 1B, 2 and 3 are all possible.
[e]Only yellow light is produced and no colour rendering is provided.
Mainly used for road lighting.
[f]Variable – consult manufacturers' literature.

light will not reveal these features. Again a simple experiment is to view an object in a conservatory and then in a room indoors. Perhaps the lighting that we find most natural corresponds to daylight through a window and so comes in at an angle of, say, 10–60° from the horizontal. Narrow beam intense light sources can create sharp-edged dark shadows (observe how, in some museums, overhead lighting causes the top part of the picture frame to shade part of the painting) For a 'softer' lighting scheme, the light needs to be diffused in some way.

A more technical discussion of this would refer to the vector/scalar ratio. Put simply, the scalar illumination is that arriving at a point irrespective of direction and the vector illumination is that from the strongest source (a detailed explanation can be found in Reference 20). The vector/scalar ratio represents the strength of the flow of light. Ratios of 1.5–2.0 give a pleasant appearance to human features; a ratio of 1.0 is a soft lighting effect and one of 3.0 gives strong contrasts.[21] Daylight can produce similar effects and if it comes in at low angles can cause objects in the space to be silhouetted. The art is again to combine daylight and artificial light to produce an enjoyable environment.

The practical importance of the above considerations is that unless daylight and artificial light work together the artificial light will not be turned off and so there will be no scope for energy savings.

Another growing concern among designers is lighting and health. SAD (seasonal affective depression) is thought to be caused by a lack of sunlight and experiments are being undertaken to see if the use of more 'natural' indoor light sources will increase the sense of well-being and health.

Light sources

The prime measure of efficiency for a light source is the ratio of the light emitted to the power input, expressed as lumens/watt. Manufacturers are constantly attempting to increase this ratio and designers are trying to keep abreast with corresponding low-energy schemes that use less power to provide the same illuminance, i.e. lumens per square metre of area lighted, expressed as lux. For office design an achievable rule-of-thumb objective is that the installation should not exceed 2.5 W/m^2/100 lux. (More detailed information is given in Reference 22).

Table 8.4 gives data on eight principal light sources used throughout the fields of domestic, commercial and industrial lighting.

Luminaires

The luminaire is the apparatus that controls the distribution of light from a lamp or lamps and that incorporates fixings and the components to connect the lamp to the power supply. One principal reason for controlling the light emitted is to avoid glare.

As has been said, glare is like intelligence – easy to recognize, but difficult to define. Essentially, glare is an imbalance between the general brightness and any particular source of light. Glare can result naturally, as when direct sunshine strikes a desk, or from artificial light sources. With glare from sunshine the danger to energy use is that blinds, or some other shading devices, will have to be used and artificial lighting increased. This 'blinds down, lights on' syndrome is by no means uncommon. It is often exacerbated by the blinds being left down even the following day which may, in fact, be cloudy.

Two kinds of glare are commonly referred to. Disability glare impairs the ability to see detail and discomfort glare causes visual discomfort.

A detailed technical discussion of designing to avoid glare can be found in Reference 24, but here we shall simply consider some broad design approaches. The standard technique for reducing glare from lamps is to incorporate some kind of diffuser or screen in the luminaire. Another technique is to recess the lamp or treat it in some other way so that most of the light falls in a fairly narrow beam. This can result in a rather gloomy effect if extreme care is not taken.

A fairly common solution that has appealed to many architects has been uplighting. Particularly in tall spaces where the ceiling has architectural interest, uplighting has been able to highlight features and provide a glareless, generally soft appearance at lower levels. The difficulty from the point of view of energy conservation is that uplighting is inherently inefficient because the reflectance from the ceiling is, at best, likely to be 85% with a bright, freshly painted white surface. Recently, designers and manufacturers have been developing combined up- and downlighters to try to improve the overall lighting efficiency and still achieve the desired aesthetic effect.

8.4 Controls

An underlying concept in thinking about controls and, indeed, buildings is zoning. Very broadly, a space will have a perimeter zone and a core zone. As the building becomes more complex it may have corridors, entirely enclosed internal spaces and atria.

Zones obviously have varying requirements. At the perimeter, control of solar radiation is most important and we have seen some solutions in Chapter 4. It is also at the perimeter that there is normally the greatest possibility of using daylight efficiently. This can be done by arranging for lights serving the perimeter to be controlled separately from others in the space. A more sophisticated (and more costly) system is to use photocells to control the lighting in a zone that can receive both daylighting and artificial lighting. As the daylight level increases the artificial lighting level is lowered (and vice versa) very gradually.

Zoning is also important in allowing for different use patterns in a space. For example, in an open plan office zoning could provide a background security level of illumination throughout the space for those working late, and local controls could increase the light levels where people were actually working.

Controls can provide a much needed (and appreciated) element of personal influence over one's working environment. In many ways this is similar to the thermal issues discussed previously. The ability to control the lighting level and direction contributes to the perception of an enjoyable environment.

One approach is to be able to dim the lights as this can help with certain glare problems as well as reduce energy consumption. Dimmers for tubular fluorescent lights have been available for some time (although they may be too costly for a number of applications) and manufacturers are striving to develop systems for other high-efficiency light sources.

A second very flexible approach is to provide background lighting to, say, 200–300 lux and then use adjustable personal task lights to increase this level as necessary. Energy efficiency is, of course, improved if these lights are turned off when they are not needed.

A technical issue for controls is how quickly light sources will respond. With several high-efficiency lamps the run-up times tend to mean that lights are left on when not required rather than switched off and switched on again when needed.

Generally, controls must be understandable. For example, switches need to be readily accessible and it should be clear what lights they serve. For many buildings, attractive schematics providing information on the controls adjacent to the switches would be very useful. Controls have been a nexus of constant conflict in the past because of conflicting requirements among individuals and between individuals and automatic control systems. 'Fiddling' with the controls is a fine art in many buildings.

More sophisticated controls incorporated in building management systems may offer reasons for optimism. Already, systems that shut off lights automatically at the end of the working day are providing energy savings, and devices such as movement detectors that ensure that lights are on in particular spaces, such as WCs, only when needed are proving useful.

One can envisage intelligent controls that 'correct' or at least warn users of the results of their decisions. For example, if a user overrides a control system that has switched off the perimeter lights on a bright sunny day, a computer could flash a message every ten minutes giving the increased energy use, CO_2 production and likely temperature rise resulting from the action. It might even, after a period of time, override the user, requiring the perimeter lights to be switched on again if desired.

The scope for control is enormous – in some buildings every single fitting is addressable and can be controlled – but, as always, the balance between simplicity and complexity will determine the overall success of any design.

Finally, any energy-efficient scheme must consider access to the controls and the lights themselves. Maintenance is a particular issue. Fittings that look slightly dirty may be wasting about 30% of their light and surveys often show greater losses. But, obviously, to clean them they must be readily accessible.

Guidelines

1. Most people prefer daylight – make effective use of it.
2. Ensure that the average daylight factor is adequate.
3. Ensure that there is sufficient uniformity of daylight.
4. Daylighting systems do not increase the amount of light – they simply redistribute it.
5. Direct sunlight from all azimuths and altitudes must be considered.

6. Consider use and maintenance.
7. Use high-efficiency fittings but ensure that they are appropriate to the application.
8. Use appropriate control systems.

References

1. Le Corbusier (1927) *Towards a New Architecture*, transl. by F. Etchells, The Architectural Press, London.
2. Anon. (1987) *Window Design: CIBSE Applications Manual*, CIBSE, London.
3. See reference 2, p. 31.
4. Anon. (1994) *CIBSE code for interior lighting*, CIBSE, London.
5. RADIANCE: Lighting programme developed at Lawrence Berkeley Laboratories and in use, for example, at De Montfort University, Leicester.
6. Anon. (1963) Estimating daylight in buildings. BRS Digest (second series) 41. BRS, Garston.
7. Anon. (1992) Lighting for Buildings: Part 2: Code of practice for daylighting. BS 8206. British Standards Institution, London.
8. Brown, G.Z., Haglund, B., Loveland, J., Reynolds, J.S., and Ubbelohde, M.S. (1992) *Inside Out: Design Procedures for Passive Environmental Technologies*, Wiley, New York.
9. See reference 2, p. 10.
10. See reference 7, p. 19.
11. Littlefair, P.J. (1990) Innovative daylighting: review of systems and evaluation methods. *Lighting Research Technology*, **22**(1), 1–17.
12. Aizelwood, M. (1993) Innovative daylighting systems. An experimental evaluation. *Lighting Research Technology*, **25**(4), 141–52.
13. See reference 11.
14. Anon. (1993) *Lighting for Offices*, CIBSE, London.
15. See reference 2, p. 47.
16. Anon. (1988) *Method of Measuring and Specifying Colour Rendering of Light Sources*, CIE Publication 13.2. Commission Internationale de l'Eclairage, Paris.
17. Anon. (1991) *Philips Lighting Catalogue*, Philips, Croydon.
18. See reference 7.
19. See reference 4, p. 28.
20. See reference 4, p. 33.
21. Turner, D.P. (ed.) (1977) *Window Glass Design Guide*, Architectural Press, London.
22. See reference 4, p. 37.
23. See reference 4.
24. See reference 4, pp. 199–205.

Engineering thermal comfort

CHAPTER 9

9.1 Introduction

Engineering thermal comfort is a broad topic. In this chapter we shall examine the range of environmental conditions encountered and look at the heating, ventilating and air-conditioning systems that are available to designers. Great emphasis is placed on natural ventilation systems because of their growing importance.

9.2 A range of conditions

Heating to cooling covers a broad field and the role of ventilation varies with the position in that field. Figure 9.1 shows the principal considerations.

9.3 Heating

Energy sources have been discussed in Chapter 7. For space heating, gas, if available, is the most common choice. Coal is not often specified at present as gas is a

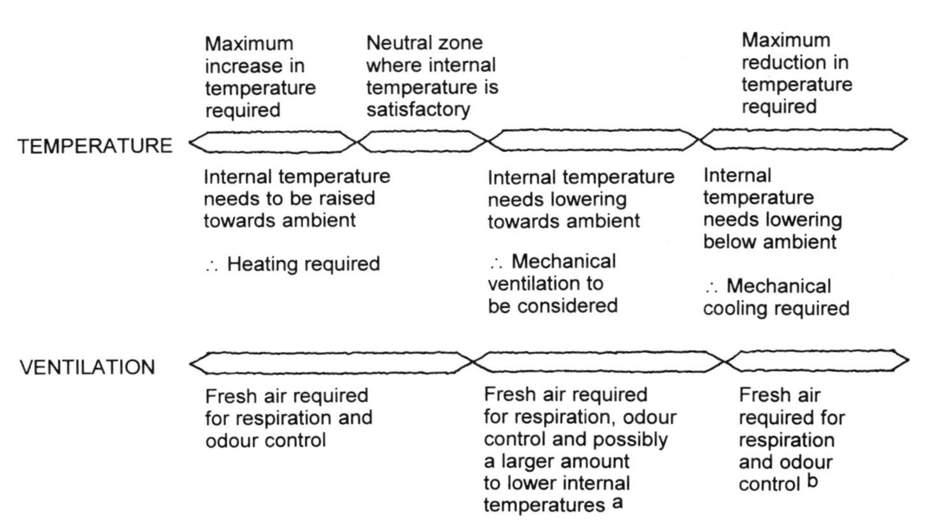

[a]This is obviously variable since, during a hot day, one may want to restrict the amount of fresh air to avoid bringing in outside air that may be warmer than the air inside.
[b]Fresh air alone, or, more commonly, a mixture of fresh and recirculated air, can be cooled below the ambient temperature. The latter is normally more energy efficient.

9.1 Heating, cooling and ventilation considerations.

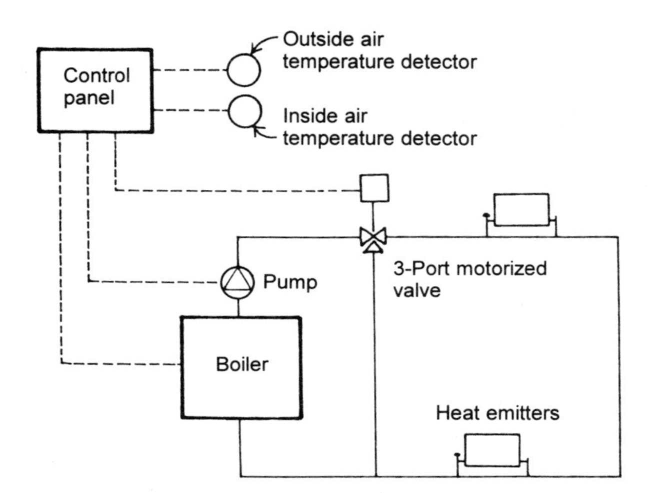

9.2 Conceptual schematic of a heating system.

cleaner option. Our discussion will therefore concentrate on gas and, to a lesser extent, electricity.

The standard 'wet' or hydronic heating system consists of one or more boilers (or combustion devices), pumps, heat emitters, interconnecting pipework and a control system, as shown in Figure 9.2.

Gas-fired boilers can be grouped into the three categories of conventional,

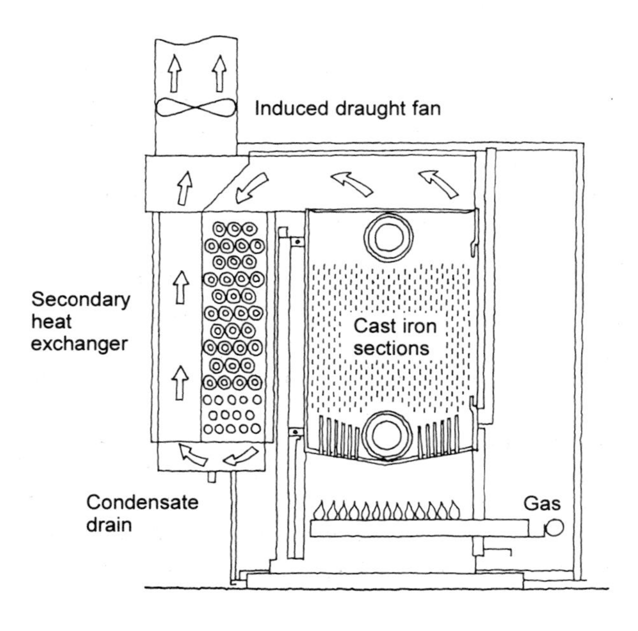

9.3 Broag gas-condensing boiler.[2]

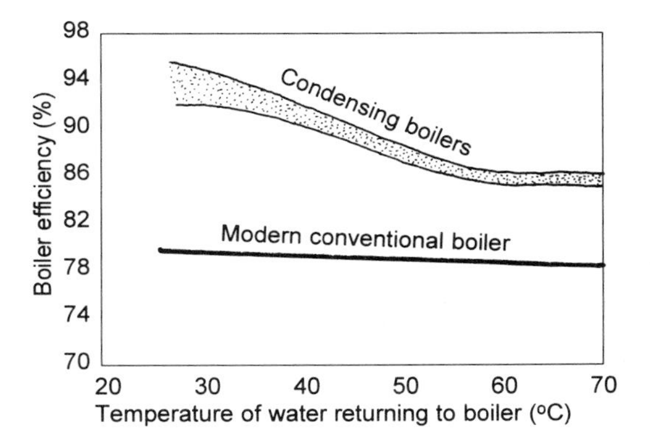

9.4 Effect of return water temperature on boiler efficiency.[4]

high-efficiency and condensing. Conventional boilers keep initial cost down by such factors as simplicity of design and reduced boiler insulation levels. High-efficiency boilers have more efficient heat exchangers and better casing insulation. Condensing boilers have an additional or enlarged heat exchanger which, under all conditions, recovers sensible heat from the flue gases (Figure 9.3) and, under suitable conditions, recovers latent heat from the condensation of water vapour (generated during combustion) in the flue gases. The formula for the combustion process for natural gas (which is principally methane) is:

$$\underset{\text{Methane}}{CH_4} + \underset{\text{Oxygen}}{2O_2} \longrightarrow \underset{\text{Carbon dioxide}}{CO_2} + \underset{\text{Water}}{2H_2O}$$

For latent heat to be recovered, the return water temperature must be below the dewpoint of the flue gases. Removal of heat from the gases results in lowered temperatures, which are typically in the range of 40–80 °C.[1]

Figure 9.4 shows the effect of return water temperature on a condensing boiler with flue gases with a dewpoint of about 54 °C. (See Reference 3 for a detailed explanation of this.) As the return water temperature falls below the dewpoint, latent heat is recovered and the boiler efficiency increases. When condensing boilers first started to become commercially available in the early 1980s, designers considered oversizing the radiators to reduce the return temperature, thus encouraging condensation of flue gases at the boiler. This was soon found to be uneconomical, however, and radiators are now sized in the normal way.

Use of the sensible heat alone in a condensing boiler gives a seasonal efficiency for the appliance of about 87% compared with a modern conventional boiler whose maximum seasonal efficiency is around 74–80%. Recovery of latent heat in the condensing boiler can increase seasonal efficiency to about 90–94%.[5] High-efficiency boilers are likely to be between, say, 80 and 87%, depending on which energy-saving features they incorporate.

When possible, high-efficiency and condensing boilers should be specified, and guidelines on design and application are given by the CIBSE.[7] Very briefly, condensing boilers are most likely to be cost-effective when above roughly 100 kW in size, when used in multi-boiler installations and when used in buildings where

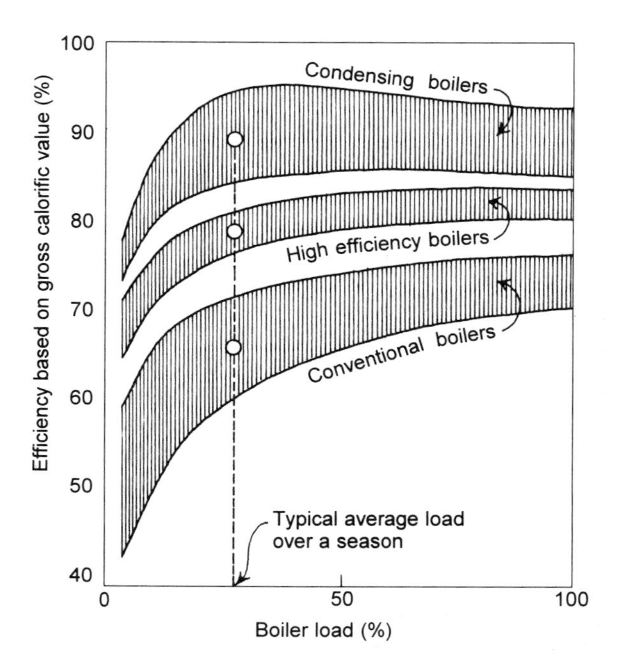

9.5 Typical efficiency ranges for boiler types at part load.[6]

the boilers must be operated for long periods. They may also be cost-effective when used as a single boiler to provide space heating and domestic hot water requirements. High-efficiency and condensing boilers are both more expensive than a conventional boiler, and for smaller domestic scale boiler sizes the costs may be significantly higher. The best way to proceed is to contact manufacturers to determine prices in a rapidly changing field. The improved efficiencies of condensing boilers do often result in very good payback periods and, of course, the reduction in energy use is accompanied by a reduction in CO_2 production.

CO_2 emissions and other pollutants can also be reduced in other ways and boiler and burner manufacturers are striving to develop appropriate products. The basic reaction for burning natural gas has been given above. Stoichiometric combustion is when the fuel is reacted with exactly the amount of oxygen to oxidize, in our example, all the carbon and hydrogen in the methane to CO_2 and H_2O. Thus in stoichiometric combustion there is no incompletely reacted fuel and no unreacted oxygen. In practice this is not attainable, and normally excess oxygen or excess air is required in the combustion process.

The more efficient the burner, the lower the amount of excess air used and the lower the volume of CO_2 emissions in the flue gases. Burners are available commercially that operate at less than 5% excess air.

Manufacturers of combustion equipment are attempting particularly to reduce pollution by lowering NOx emissions (Chapter 2). NOx are formed during combustion by reactions of nitrogen and oxygen at high temperatures and by oxidation of organic nitrogen in (certain) fuel molecules.[8] Part of the effect is related to excess air because NOx levels peak when excess air is at 40% and reduce sig-

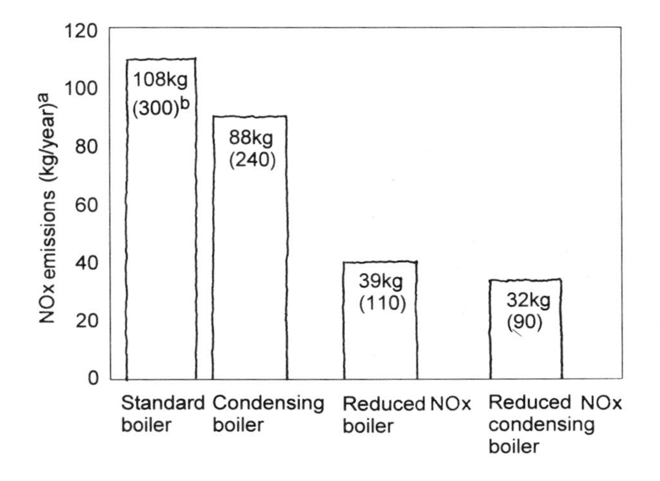

[a]Based on a 300 kW boiler operating for 1200 hours per year.
[b]Figures in parentheses are mg NOx per kWh of delivered energy.

9.6 NOx emissions from different boiler types, with and without reduced NOx burners.[11]

nificantly as the excess air decreases.[9] The main part, however, concerns control of the burner flame temperature. Methods of reducing NOx emissions include burner adjustment, flame inserts and staged combustion and delayed mixing.[10]

A bewildering range of units exists for expressing NOx emissions and care must be taken in comparing different manufacturers' products. A convenient unit is mg NOx per kWh of delivered energy. Figure 9.6 indicates some typical NOx emissions from different types of boilers.

We have seen in Chapter 2 that SOx (SO_2 and SO_3) are air pollutants. SO_2 is produced when sulphur or sulphur-containing fuels (such as coal) are burned. As SO_2 is the result of the fuel, sophisticated burner techniques have little effect on reducing its formation. SO_3 is created in the atmosphere by the oxidation of SO_2 under the influence of sunlight; some can also be produced directly in the combustion process.

It is interesting to note that clients with important building portfolios are now including in their environmental policy statements specific directives to designers to minimize SO_2 and NOx emissions to the atmosphere when designing and installing new equipment. (For existing plant the same clients require a high standard of maintenance to help reduce these emissions.)

9.4 Heating: distribution systems and heat emitters

Hydronic heating systems (i.e. those containing water) with gas-fired boilers are the most commonly installed. Pipework tends to be steel in larger installations and copper in smaller and domestic buildings. Energy losses associated with distribution can be reduced by locating the energy sources near the centre of the load. Careful attention to insulation of pipework – ensuring that energy is delivered only at the emitters – is another way to lower energy consumption. A range of typical emitters and their approximate net outputs is shown in Figure 9.7.

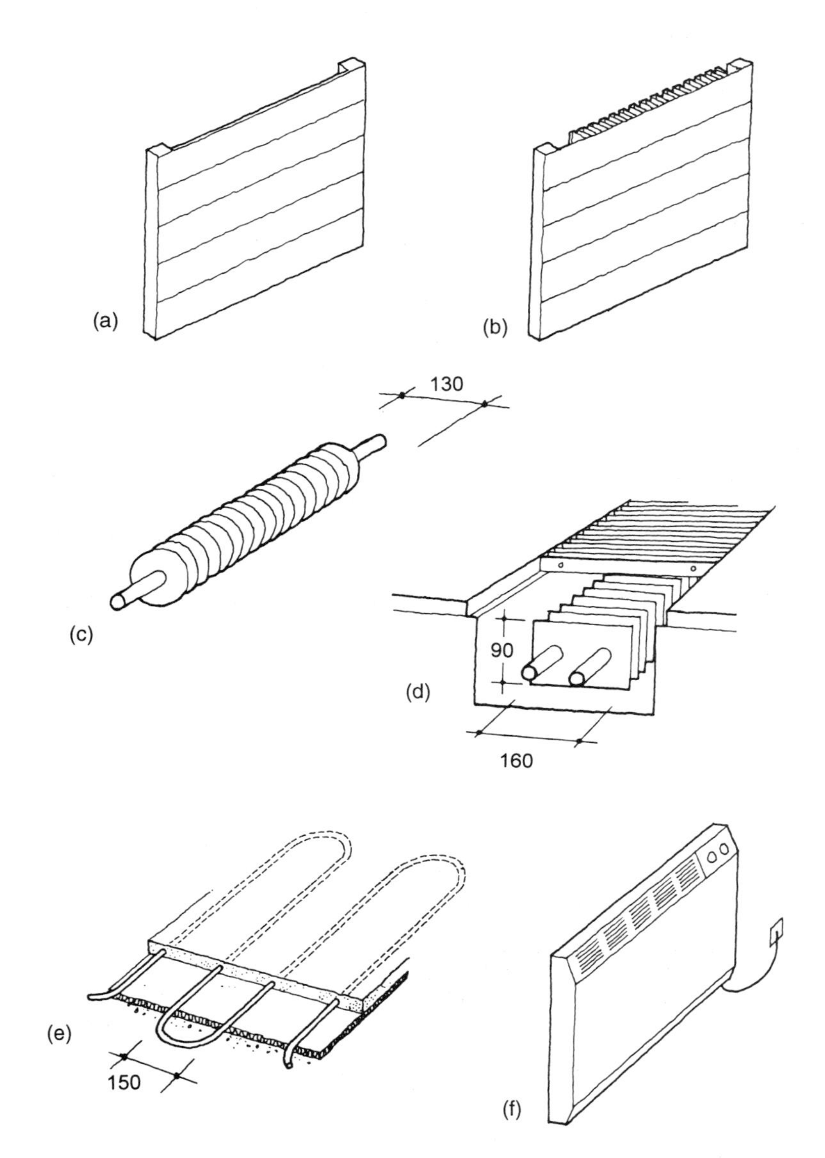

(a) Panel radiators *ca.* 1300 W/m^2
(b) Panel radiators with fins *ca.* 2500 W/m^2
(c) Spiral-gilled tubes *ca.* 600 W/m^2
(d) Perimeter trench heater *ca.* 1200 W/m^2
(e) Underfloor heating *ca.* 90 W/m^2
(f) Electric convector heater *ca.* 5600 W/m^2

9.7 Typical heat emitters and approximate heat outputs.

The choice of which emitter to use is often related to issues of appearance and space. In terms of energy efficiency an important consideration is the pattern of occupation of the building. The thermal response of the building fabric is a vital consideration also. For intermittently occupied spaces a rapid thermal response is required from the heat emitters so that the occupants do not have to wait long to be comfortable. In continuously occupied buildings this is less important. Another consideration is that if casual heat inputs from, say, solar gains or a crowd of people increase quickly, one wants the heat emitters to respond by

Table 9.1 Approximate thermal response of heat emission methods

Thermal response category	*Heat emitters*
Rapid	Forced warm air, unit heaters, fan convectors
Intermediate	Natural convectors, radiators, perimeter heaters, spirally wound tubes
Slow	Underfloor heating

decreasing their outputs to avoid overheating. There is thus an interaction between heat emitters and their controls. Table 9.1 gives an approximate grouping of the thermal response of some heat emitters.

Warm air heating systems have been in common use in American homes for many years. Because suburban bungalows often have basements, ductwork can easily be run between basement ceiling joists and taken up to wall registers that are manually controlled. In the UK, with space at a premium, fewer warm air systems have been installed, either in domestic or commercial buildings.

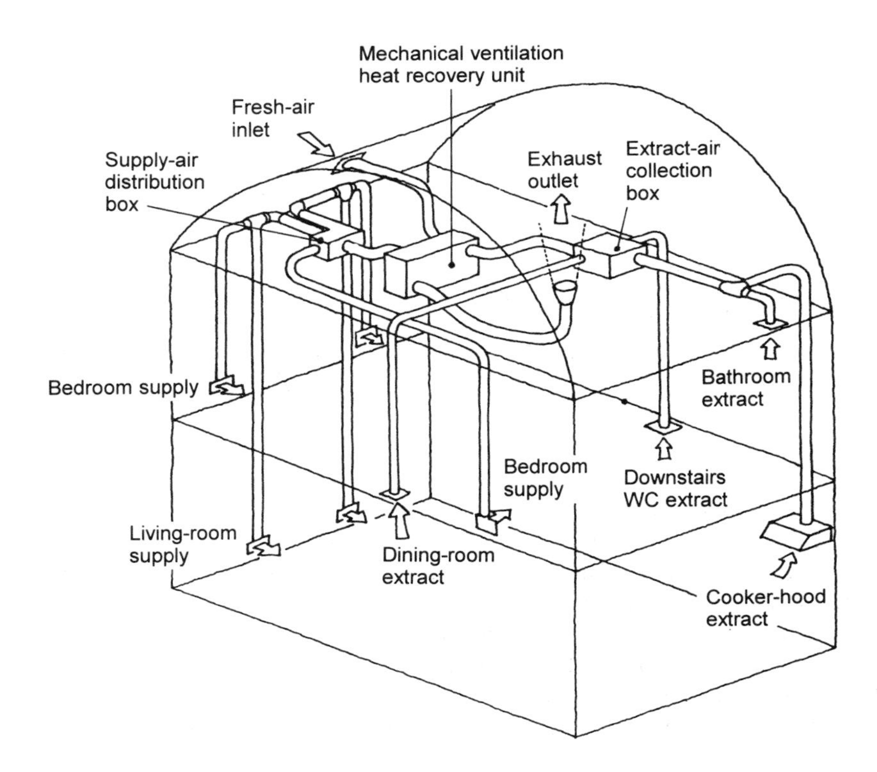

9.8 Domestic warm air heating system.

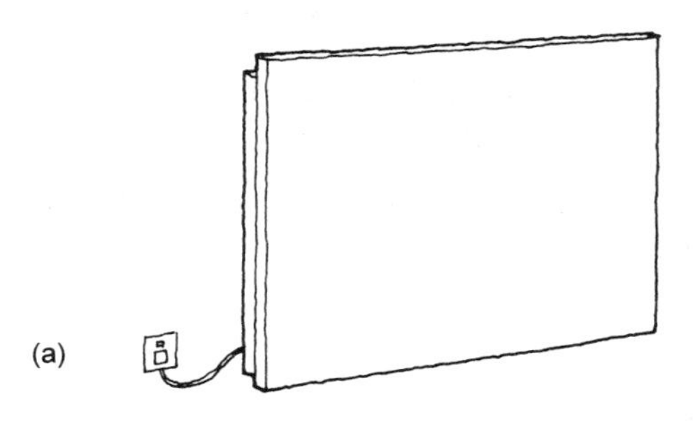

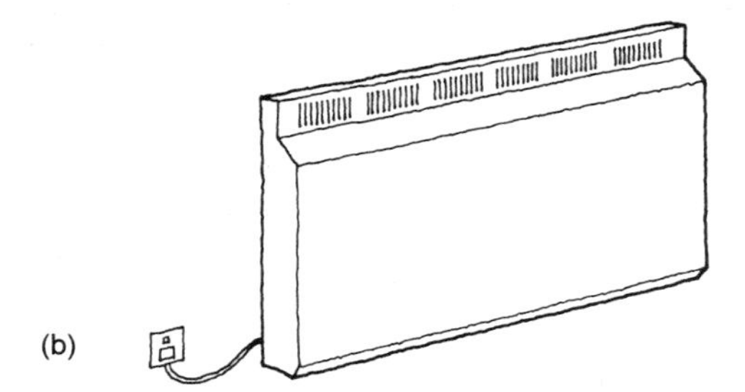

(a) Panel heater
(b) Storage heater

9.9 Electric heaters.

Interest has grown of late because of the possibility of incorporating heat recovery on the extract air. This has occurred simultaneously with improved building insulation and so the ventilation heat loss has become more important in relation to the fabric loss. Figure 9.8 shows a typical domestic warm air heating installation. Ventilation and heat recovery are discussed in more detail below.

Electric heating from the present mix of fuels feeding the national grid is not the best environmental choice (Chapter 7). Nonetheless, in particular circumstances electric heaters are used and a range of emitters is available. Instantaneous heaters include the old-fashioned radiant bar elements, electric panel heaters (Figure 9.9) and compact heaters with fan assistance to increase heat output.

For some time the electricity supply industry has argued (very reasonably) that one must look at the building and heating system together. The industry has pointed to the lower installation and maintenance costs associated with electricity and argued that by using some of these savings to improve the insulation levels of buildings, electricity use is decreased to a point where electric heating can be cost-effective. This is particularly true, it is claimed, if part of the supply is at off-peak rates through the use of storage heaters. Of course it can be argued that the insulation level of all buildings should be to quite a high standard to begin with, and the next step is to examine what the fuel source and heating system should be.

A recent study by the Association for the Conservation of Energy of all-electric houses concluded that the 'economics are dubious at best' and that even when built to a higher-than-normal standard (i.e. their Medallion 2000 homes as opposed to those built to 1990 Building Regulations) energy use was double and emissions nearly four times as much as a gas-heated house.[13]

A study of electric heating at the BRE office building[14] (where panel heaters are used and there is no mechanical ventilation system) included the following points:

1. Electrical heating systems offer simplicity of design and installation and potentially reduced capital costs, low maintenance and increased flexibility.
2. All electric direct-acting heating systems have the potential for good local temperature control.
3. A comprehensive economic evaluation of the various services and fabric options needs to be carried out beforehand for each particular building project.
4. Carbon dioxide, nitrous oxide and sulphur dioxide emissions are higher with electric heating than with gas heating. The amount by which they are higher depends on a number of factors, including the insulation standard adopted for the building and the relative efficiencies of the two systems in delivering useful heat.

If we compare the embodied energy in the two alternative systems, the embodied energy and CO_2 audit of the BRE Low Energy Office showed that gas-fired systems have about five to six times the embodied energy of electric systems; however, the additional energy and thus, again, CO_2 is paid back in about a year because energy and emissions in use are so much less.[15]

9.5 Ventilation

The requirements for ventilation have been described in Chapter 2. In large buildings, ventilation played a number of roles. The nineteenth- and twentieth-century urban atmosphere was often extremely polluted by the products of combustion, especially from coal, and engineers and architects of the time began to investigate ways of producing cleaner air inside their buildings.

One of the landmarks of twentieth-century modernism, Frank Lloyd Wright's Larkin Building (Figure 9.10), was designed as a sealed box with mechanical ventilation to combat pollution (and probably noise) from the adjacent railway.

The development of highly mechanically serviced buildings in the twentieth century can be explained by a number of factors, foremost among which was cheap energy – the running costs for buildings were not sufficiently high to encourage investigation of alternative ways of servicing buildings. (To an extent this is still true and the major impetus at present for more progressive policies is environmental, i.e. the greenhouse effect and depletion of the ozone layer rather than cost.) The pristine glass block was also a potent symbol of architectural modernity and social progress. These buildings were seen as prestigious, and an alliance of clients, architects, structural engineers, mechanical engineers and suppliers of everything from ventilation fans to cladding systems collaborated to produce them. In many ways these systems were effective. Filters allowed them to remove many outdoor pollutants, ductwork systems with attenuators were able to reduce external noise significantly and cooling permitted comfortable temperatures to be maintained in the hottest conditions. The price has proved to be too

9.10 Larkin Building by Frank Lloyd Wright.[16]

high, however, and we need to determine the most appropriate way to provide comfortable temperatures while simultaneously addressing related issues such as air quality and noise.

A possible starting point is vernacular architecture. What kinds of solutions were developed from generations of practical experience and immediate observations of the effects of design variations? Figures 9.11 and 9.12 show details from English maltings; Figure 9.13 shows a Scottish whisky maltings; and Figure 9.14 a nineteenth-century American home with a cupola (also known as a lantern, belvedere, observatory and widow's walk).

9.11 Maltings kiln vents at Langley, Worcestershire.[17]

9.12 Maltings air inlet detail at Mistley, Essex.[18]

9.13 Scottish whisky maltings at Dalwhinnie.[19]

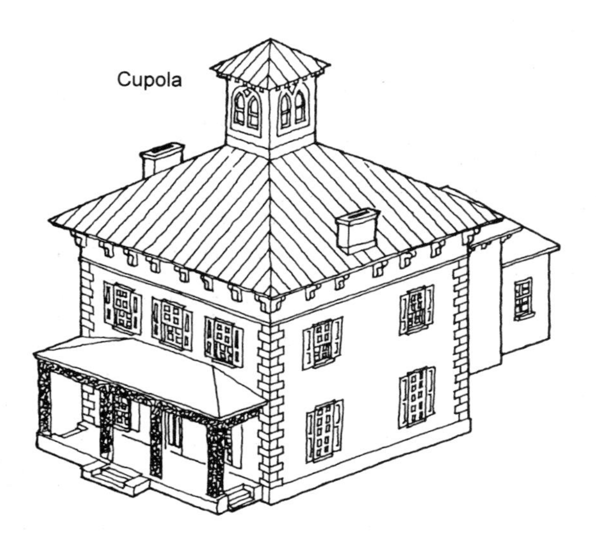

9.14 Nineteenth-century American home.[20]

These buildings first of all strike us by their character and point us towards an architecture that can combine functionalism, fantasy and a sense of place. If we look at them more technically we note that the areas for air to enter and leave are large. In the maltings air comes in through the large louvred windows and passes up through the flooring blocks, then out through the kiln-vent. (Temperatures are, of course, higher than most of us would find comfortable.) In the American house the cupola was centred over the staircase and by opening windows and internal doors on all levels ventilation occurred from side to side and bottom to top. Protection against rain is provided by louvres in the kiln-vent in the maltings and windows in the house. Air paths are simple and direct.

If we now consider our naturally ventilated homes we know that they are usually ventilated from one side (Path 1) or cross-ventilated (i.e. ventilated from one side to the other) (Path 2). Figure 9.15 shows this and additionally has a third path which is stack effect ventilation.

The natural driving forces for ventilation are the wind and stack effect. These are, of course, variable and less easy to control than the forces associated with mechanical systems. On the other hand, where used successfully they can enormously reduce capital and running costs of mechanical and electrical plant and reduce the need for plant space. Figure 9.16 shows indicative flow patterns over a simple building.

The driving force of the wind, and hence its ability to ventilate a building, is related to its velocity pressure, VP, which is given by:

$$\mathrm{VP} = 0.63\, u^2$$

where VP is in pascals (Pa) and u is the speed of the wind (m/s).[22] Thus if the wind speed is 2 m/s, the velocity pressure is about 2.5 Pa and if the speed is doubled, the velocity pressure becomes 10.5 Pa. These pressures are very low

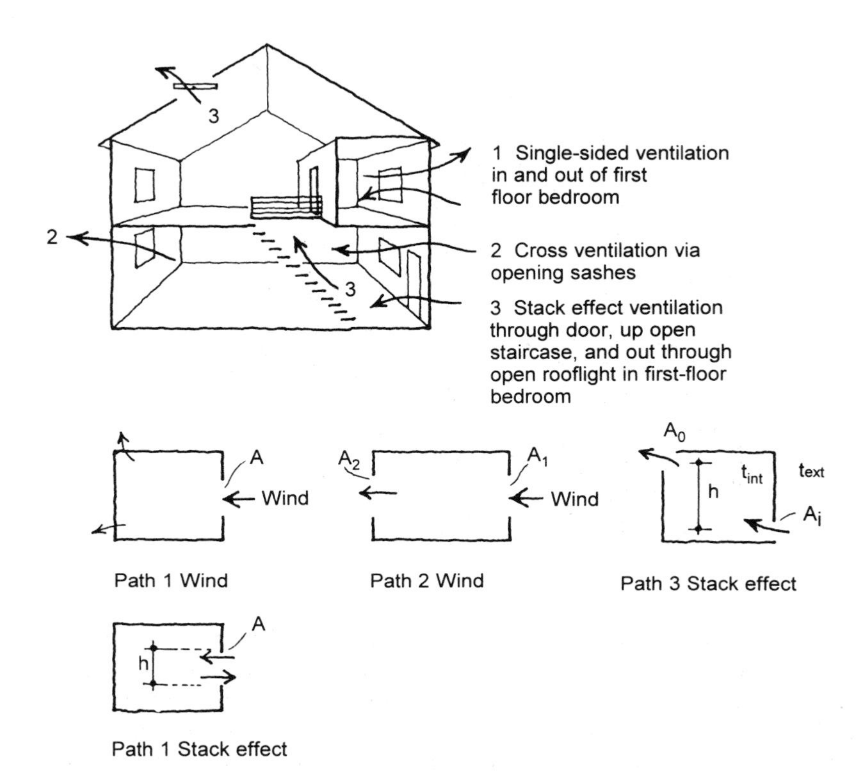

9.15 Ventilation paths in a simplified and modified Victorian terraced house.

compared to those encountered in mechanical ventilation systems.

Very approximately the pressure on the windward face of a building is about 0.5–0.8 times the velocity pressure of the wind, and on the leeward side the negative pressure is about 0.3–0.4 times the same pressure. The total wind pressure acting across a building is thus approximately equal to the wind velocity pressure. (More detailed discussions of these issues are found in Reference 22.)

The wind speed at Heathrow Airport in London exceeds 1.6 m/s about 80% of the time; at lower heights the wind speed is reduced and in more urban areas it is further reduced. The conclusion is that in many situations wind cannot be relied upon to provide any significant driving force. (The coincidence of a critical situation of low wind speeds and high temperatures is discussed in Appendix A.) Another point to note is that because the wind speed is so variable, the ventilation rate of naturally ventilated buildings changes more than in a tightly controlled mechanically ventilated construction.

Returning to Figure 9.15, let us examine each path quantitatively.

Path 1 The amount of ventilation due to the wind alone is given by:[23]

$$Q = 0.025\, Au$$

where Q is the volume flow rate (m^3/s), A is the area of the opening (m^2) and u is the wind speed (m/s).

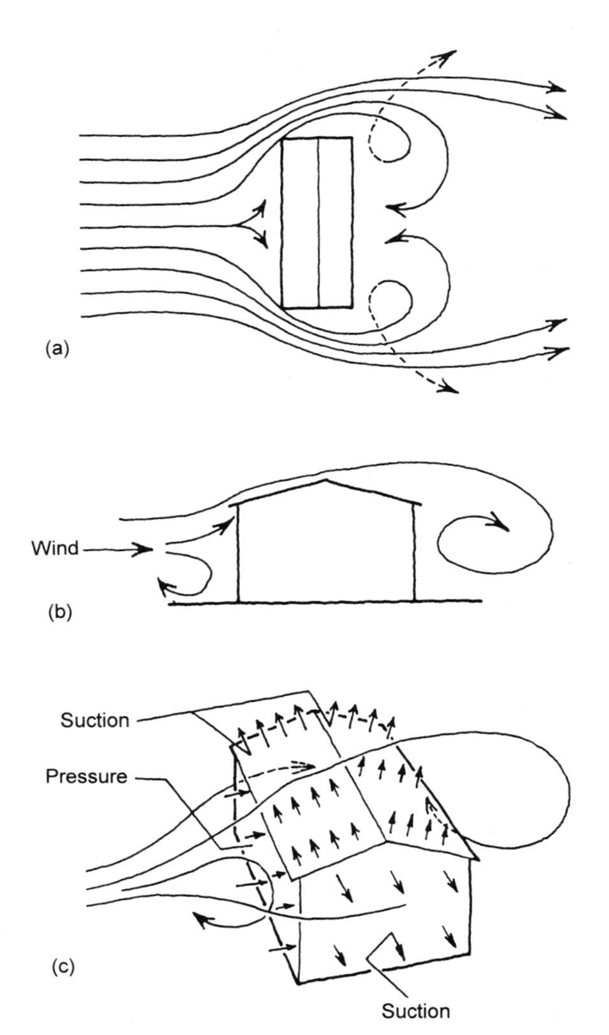

9.16 Schematics of air flows and resulting pressures on an isolated building.[21]

Assuming $u = 2$ m/s and $A = 1$ m^2, we have

$$Q = 0.05 \text{ m}^3/\text{s}$$

Path 1 can also have a stack effect. The term refers to the process of heated air (which is, of course, less dense) rising in a space. Air inside most of our occupied buildings tends to be warmer than outside air. It rises, finds its way out through an available opening and is replaced by cooler external air at low level.

The stack effect equation is rather more complicated:[24]

$$Q = 0.2A \left[\frac{(t_{int} + 273) - (t_{ext} + 273)}{t_{mean} + 273} gh \right]^{0.5}$$

where t_{int} is the internal temperature (°C), t_{ext} is the external temperature (°C), t_{mean} is the mean of t_{int} and t_{ext} (°C), g is the gravitational constant (9.81 m/s^2) and h is the height of the opening (m). (This formula assumes a discharge coefficient for flow through the window of 0.61.)

Assume $A = 1\ m^2$, $t_{int} = 20\ °C$, $t_{ext} = 15\ °C$ and $h = 1$ m and we have

$$Q = 0.08\ m^3/s.$$

Path 2 Cross ventilation by the wind is given by

$$Q = 0.61\ A_w u\ (\Delta C_p)^{1/2}$$

where A_w is the equivalent area (m^2) for wind-driven ventilation, given by:

$$\frac{1}{A_w^2} = \frac{1}{A_1^2} + \frac{1}{A_2^2}$$

and ΔC_p is the change in the pressure coefficient. ΔC_p ranges from 0.1 for a sheltered site to 1.0 for an exposed one. Again we have assumed a discharge coefficient through the window of 0.61.

If we take A_1 as 1 m^2 and A_2 as 1 m^2, u as 2 m/s and $\Delta C_p = 0.5$, we have

$$Q = 0.62\ m^3/s$$

This shows the commonly appreciated advantage of cross ventilation over single-sided ventilation.

There is also stack effect ventilation associated with Path 2 but we shall not cover it here.

If both wind and stack effects operate at the same time, the behaviour is complex and depends on the relative values of the opening areas and the relative values of the wind and stack effects. A reasonable approximation of the combined flow rate is simply to take the larger of the two separate values. Note that the resultant ventilation is not the sum of the two.

Path 3 Here we see a more pronounced stack effect ventilation path. The basic pressure difference equation for stack effect ventilation is given in a simple form by:[25]

$$\Delta p = 3462h \left(\frac{1}{t_{ext} + 273} - \frac{1}{t_{int} + 273} \right) \text{Pa}$$

where h is the vertical distance between the inlet and outlet (m), t_{ext} is the external temperature (°C) and t_{int} is the internal temperature (°C). For common temperatures encountered in building design an approximate form of this equation is:

$$\Delta p = 0.043h\ (t_{int} - t_{ext})\ \text{Pa}.$$

Thus, if h is 6 m and the temperature difference is 5 °C, the pressure difference is 1.3 Pa. Again, this pressure difference is very low compared to those encountered in mechanical systems.

The volume flow rate is given by:

$$Q = 0.827\ A\ (\Delta p)^{0.5}$$

where

$$A = \frac{A_i \quad A_o}{(A_i^2 + A_o^2)^{0.5}}$$

in which A_i is the area of the inlet and A_o is the area of the outlet.

Assuming $A_i = 1.6\ m^2$ for the door in our case, and A_o is the same for the roof light, then

$$Q = 0.93\ m^3/s.$$

If the volume of the house in Figure 9.15 is 200 m^3, the hourly air change rate would be about 17. Try comparing the above calculations with the ventilation of your own home in summer.

Note that the stack effect is proportional to the square root of the temperature difference and that the temperature difference is greater in the winter than in the summer. This is unfortunate in a sense because the driving force is greatest when it is least needed and least when it is most needed.

In the design of office buildings the forces acting on the building are, of course, similar to those on a house. As the loadings due to solar gains, occupants and equipment and the form of construction will, however, vary enormously, it is difficult to generalize, but as a very rough guideline single-sided ventilation is thought to be useful up to about 6 m.[26] Cross ventilation is thought to be effective up to above five times the height of the space; thus, for a 3 m floor-to-ceiling height the window-to-window distance might be 15 m.[27] In the author's offices – which are essentially circular in form (with a diameter of 23 m), have an average height of about 3.5 m and are open plan – cross ventilation from 22 windows located around the perimeter is adequate.

Recently, the BRE[28] has suggested that single-sided ventilation is effective up to about 10 m, based on tests of a south-facing room 10 m deep, 6 m wide and 3 m high; about 18% of the south faade was openable glazing. These findings require corroboration in practical studies and, for the present, it is probably best to consider 6 m as a maximum.

Where single-sided ventilation is not sufficient (or where noise limits its use) and where cross ventilation is not possible due to, say, compartmentation of spaces, stack effect ventilation may be a possibility.

Detailed knowledge of the effectiveness of these systems is limited but guidelines are being developed on the basis of theoretical studies and practical experience. For example, a design guide for naturally ventilated courtrooms[29] recommends the following minimum figures:

courtroom height	= 6 m
stack height	= 5 m (*NB*: in this case the outlet air position could be below ceiling level)
free area of low-level inlet	= 1% of floor area
free area of high-level outlets	= 2% of floor area

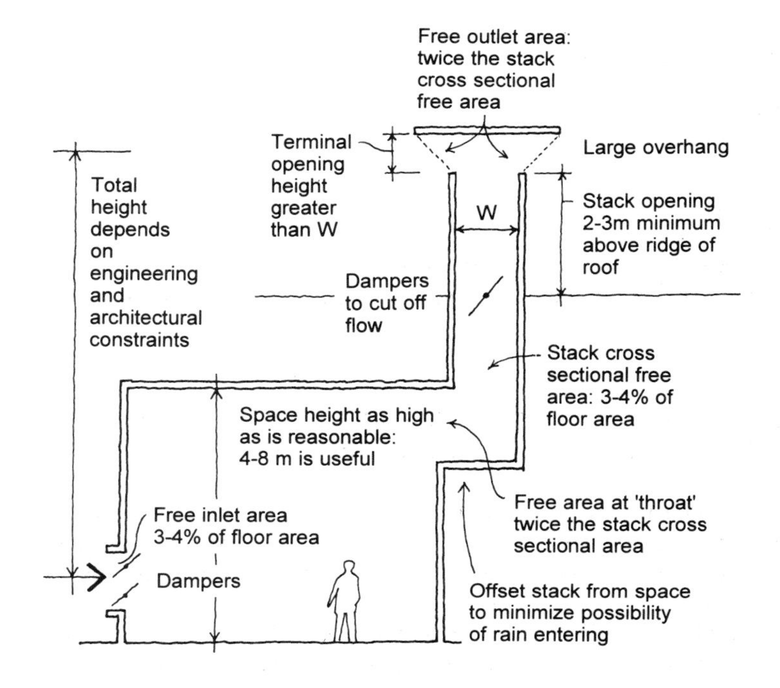

9.17 Simple stack effect ventilation.

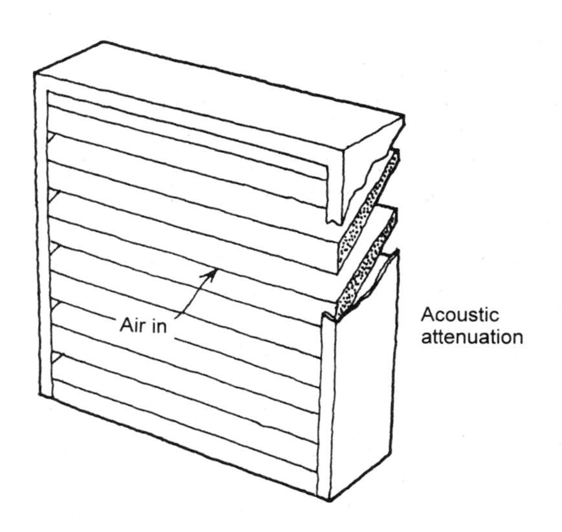

9.18 Louvre design.

At the De Montfort University Queens Building auditoria (Chapter 13) the as-built comparable figures are:

auditorium average height	= 5.1 m
stack height	= 12.0 m
free area of low-level inlets	= 4.8% of floor area
free area of high-level outlets	= 4.8% of floor area

Table 9.2 Checklist for naturally ventilated spaces

1. *Construction and configuration of space*
2. *Ventilation and expected internal temperatures*
   A. Stack effect
   B. Wind assistance
   C. Mechanical assistance
   D. Comfort and avoidance of overheating
   E. Room air movement
3. *Supply air*
   A. Source
   B. Tempering and heating generally
   C. Bird screens
4. *Extract air*
   A. Normal path
   B. Possibility of heat recovery
   C. Bird screens
5. *Smoke ventilation and fire prevention*
6. *Avoidance of rain and wind-driven snow*
   A. Prevention
   B. Draining if prevention fails
7. *Lighting*
   A. Daylighting and blackout facilities
   Interference with ventilation path
   B. Artificial
8. *Acoustics*
   A. Acceptable noise levels
   B. Room acoustics
9. *Risk of condensation*
   A. Interstitial
   B. On any part of exposed structure
10. *Controls*
11. *Durability, access and maintenance*
12. *Appearance*
13. *Cost*

Notes:
1. Naturally ventilated systems are not easily compatible with provision of incoming air filtration because of the additional resistance that results.
2. Wind loads need to be assessed by a structural engineer.

Considerably more work is required before consistent and detailed guidance can be provided for a range of building types and a range of loadings from the Sun, occupants and equipment.

Figure 9.17 summarizes some rules of thumb and design aspects that may prove useful for preliminary designs.

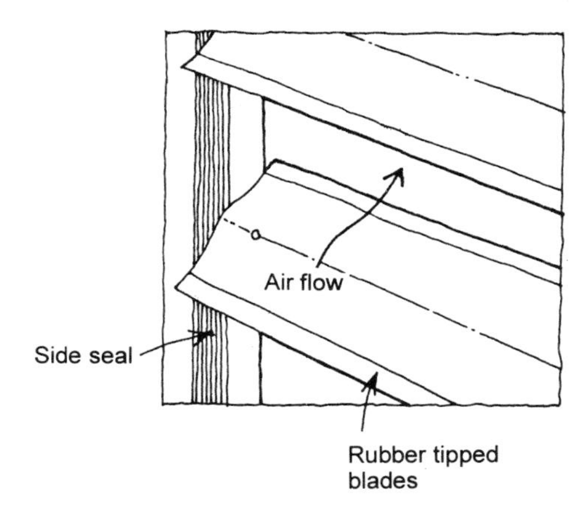

9.19 Damper detail.

The procedure that is followed in developing real designs is, of course, more complicated. Essentially what is done is to assume an acceptable peak temperature in the space based on comfort considerations and discussions with the client, which thus gives the maximum temperature differential available to drive the stack effect. (This might be 2 °C above the ambient temperature.) Openings are then sized to give a ventilation rate. The ventilation rate is then put into a peak summer-time temperature calculation (using the admittance procedure, which takes into account the heat loads in the space and numerous other factors)[30] to see if the initial peak temperature and opening areas are suitable. The procedure is carried out iteratively until agreement is reached.

To transform Figure 9.17 into architecture is a challenge, and Table 9.2 lists some of the considerations; unfortunately, lack of space precludes anything other than a brief mention of one or two of them.

At the air inlet some kind of louvre may be used to allow air in and keep birds out while not causing too high a pressure drop. Figure 9.18 shows typical louvres with a free area of about 60%.

Louvres similar to those in Figure 9.18 have been used for some time to allow air entry to boiler rooms, as air inlets to mechanical ventilation systems and in acoustic enclosures. For natural ventilation systems their pressure drop must be examined closely.

As we have seen in Chapter 4, acoustics must also be treated with great care. Mechanical ventilation systems owe their past success as much to their ability to deal with external noise as with temperature. Acoustic louvres on their own may only be a starting point. If more attenuation is required it will be necessary to introduce a path with a low pressure drop. (Chapter 13 briefly discusses how this was dealt with at De Montfort University. As can be seen there, the removal of noise usually requires space!)

Behind the louvres will be dampers that will probably be automatically controlled to shut off the air flow when it is not desired, such as when the space is unoccupied at night in winter. Figure 9.19 shows a detail of dampers with a free area of about 90% when open, and metallic side seals and blades tipped with rubber to minimize air flow through them when closed. When open the dampers must not, of course, have a high resistance to air flow.

Behind the dampers at some point there must be a heat emitter to ensure that before the air enters the occupied zone in winter it is 'tempered', i.e. brought up to a reasonable temperature that will avoid the discomfort associated with cold air movement. As with previous components, the heat emitter must not have a high resistance to air flow. The issue of draughts and air movement is important. If air is delivered under seats or close to people, possible discomfort must be considered. One approach is to deliver air at high level and let it mix with room air before it reaches the occupied zone. This can have the advantage of not causing high room air speeds at desk level in offices, as air movements of 0.5 m/s can move papers. The situation requires careful examination, however, since, with some room geometries and ventilation patterns, the introduction of air at high level could displace a layer of hot stratified air downwards towards the occupants. In this respect, specialist advice may be needed.

In situations where noise is not an issue, the combination of louvre and damper can be dealt with by an open window that can be manually or automatically controlled. Windows must meet a variety of, at times, conflicting requirements in natural ventilation systems, and their design is not easy. They need to be easily openable and easily secured and any associated blinds or shades must also be easily operable.

Table 9.3 provides some thoughts on window types. Note that windows may also be used as the air outlets. Obviously some of the considerations listed apply to single-sided and cross ventilation systems as well.

The air outlets form an area where designs will be refined over the next few years. If we first consider areas, the free outlet of the stack should be about the same as the free inlet area, judging by studies for De Montfort University's Queens Building. The free area of the termination (sometimes called the monitor) should be about twice the free area of the stack. Again, the guiding rule is always to minimize the resistance to air flow. It is very desirable to have the termination open on all sides so that when it is windy, winds from all directions can assist the stack effect. For the same reason it is important to raise the termination above the ridge or any obstruction to remove it from the more turbulent zone created by wind passing over a building.

A wide variety of designs for tops of stacks can be made to work and so there is considerable scope for architectural expression. Solutions vary from the maltings type of Figure 9.11 to those of De Montfort (Chapter 13) to the new lantern of St John's College, Cambridge, shown in Figure 9.20, which both admits light and exhausts air.

At St John's the areas ventilated by the lantern have doors that are held open magnetically and close on a fire signal. Generally, a great deal of attention needs to be given to smoke ventilation in assisted naturally ventilated buildings. Escape routes require careful consideration. There is also an alternative approach to that of St John's whereby buildings are compartmentalized and each compartment has its own ventilation system and fire control strategy (for example, see Figure 9.26).

Table 9.3 Windows for ventilation systems

Type	*Comments*
A. Bottom hung with glass sided hopper 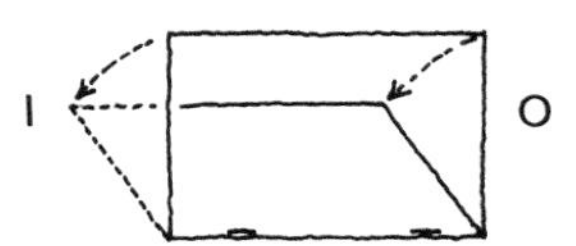	Common in nineteenth century design. Keeps rain out. If left open can be used for night-time ventilation while still providing security. Disadvantage is that there is a reduction in the free area of the window and so flow through is limited.
B. Top hung 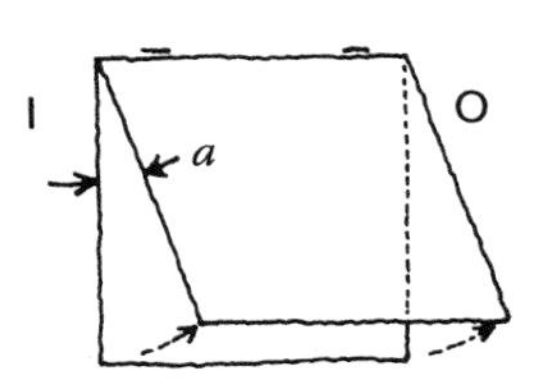	Can be easily motorized. Restriction to air flow is a function of the angle of opening (CD is the discharge coefficient, i.e. a measure of the ease of air flow through an opening) *a* — *CD*[31] 30° — 0.42 60° — 0.57 90° — 0.62
C. Centre pivoted 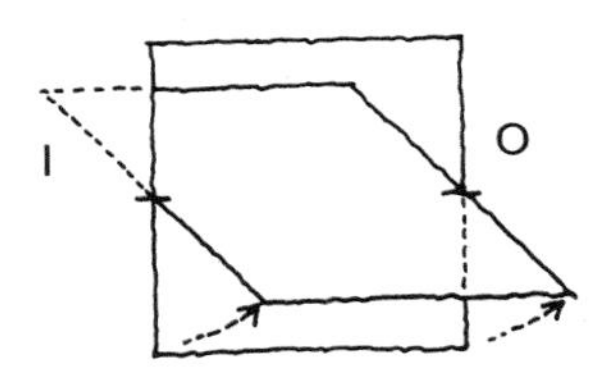	May impinge on circulation space. Restriction to air flow is a function of the angle of opening.
D. Side hung 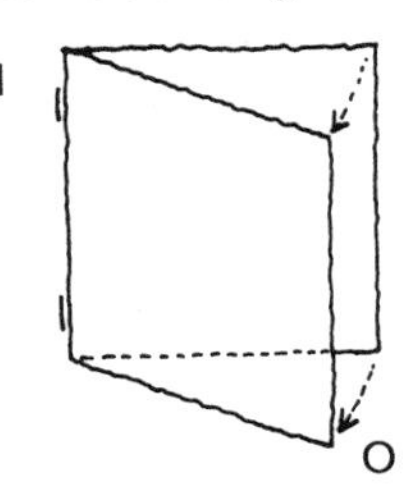	Leaves a large unimpeded area for stack effect ventilation but may block wind driven ventilation. Automatic opening of these windows may be difficult.
E. Sash 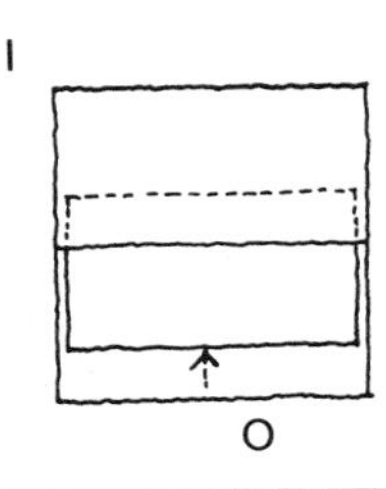	Allows variable amounts of ventilation at top and bottom. Does not protrude into space. Crack leakage can be difficult to control (as for all types of sliding windows)

I = inside, O = outside.

9.20 St John's College Library, Cambridge.

It may prove of some benefit to make two final points on design:

1. Constant attention to detail is required when developing natural ventilation systems. The clash of a rainwater downpipe or a structural support with the air path, or spacing behind seats which creates draughts or inadequate controls and so on, are more likely to jeopardize success than with mechanical solutions.
2. On a less serious note, our rule of thumb is that the resistance of the air path is directly proportional to the number of designers and the length of the design process.

To conclude this discussion it cannot be overemphasized that much more work is required in this field in order to optimize the type of system described above.

Infiltration

Let us consider for a moment the opposite problem of too much uncontrolled air infiltrating the building when it is not required in the winter and thus increasing the heating demand.

In the past, air has got in through a variety of cracks in the building such as those around door and window frames, junctions between roofs and walls or floors and walls and joist penetrations (Chapter 4). The amount of air getting through cracks or small openings is given by

$$Q = kL\,(\Delta p)^n$$

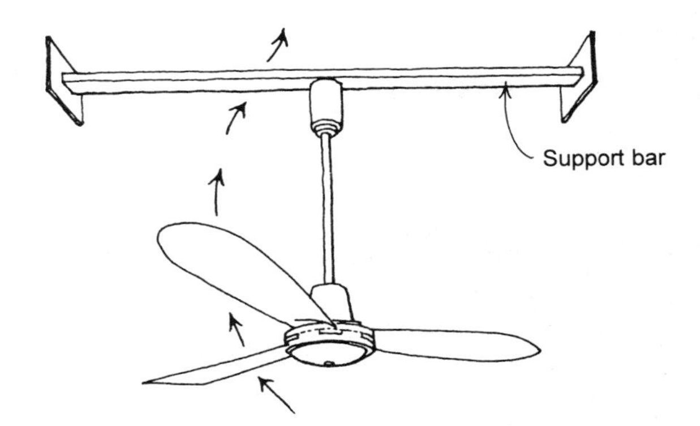

9.21 Propeller fan.

where Q is the flow rate (m^3/s), k is the window leakage factor, L is the length of crack (m), Δp is the applied pressure difference (Pa) and n is the flow exponent, taken as 0.67. Typical values of k in litres per second per metre of crack length for an applied pressure difference of 1 Pa are 0.08 for sliding windows and 0.21 for pivoted windows.[32]

Because a small opening can allow significant amounts of air to enter, in buildings designed for summer-time natural ventilation great care is needed to ensure that the air paths can be effectively sealed when not in use, i.e. particularly at night during the winter. All components in the air path involved in controlling the flow, such as dampers, windows and flaps, must be capable of being tightly sealed. If this is not achieved, the buildings may well be draughty and have high winter-time heat losses.

Assisted natural ventilation systems

If, moving towards the right of Figure 9.1, additional ventilation is required, it may be possible to add a simple fan to the system. This might take the form of a three-bladed propeller (or punkah fan), as shown in Figure 9.21.

Performance data for these fans is not always readily available. Our measurements for a typical 1350 mm diameter fan installed in a spacious vertical duct of low resistance (and so approaching 'free space') indicated a flow rate of about 0.7 m^3/s when the fan was running quietly at about 80 rev/min.

Assisted natural systems have the great advantages of increasing the designer's (and client's) confidence that the system will cope with peak conditions. They also allow greater control and can be particularly advantageous for night-time cooling. Research is needed on the optimal mix of natural ventilation paths (which, of course, take up space and require construction materials) and simple mechanical fans. There is nothing sacred about 'pure' natural ventilation, and assisted natural ventilation systems are likely to provide the best way forward.

In theory it is possible to incorporate heat recovery in these systems because of the additional driving force provided by the fan, but in practice it may be more difficult. The resistance of the fan in the airstream may be excessive, the pressure

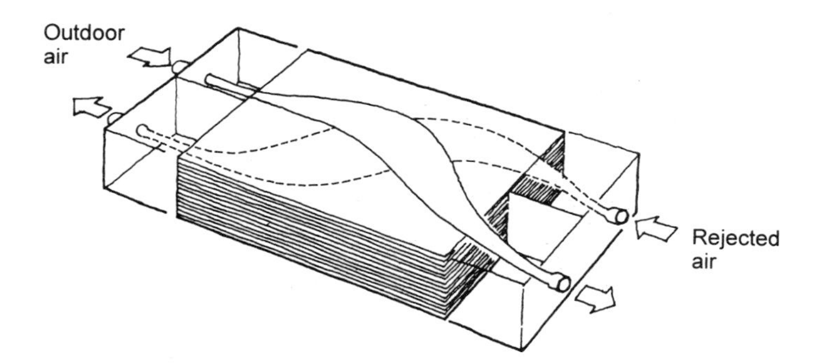

(a) Cross-flow heat exchanger

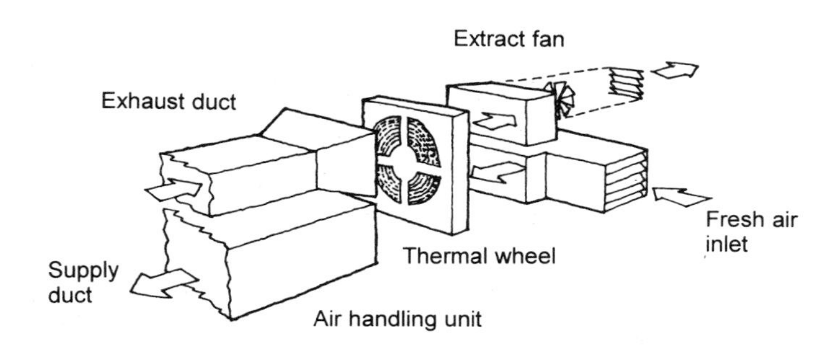

(b) Thermal wheel

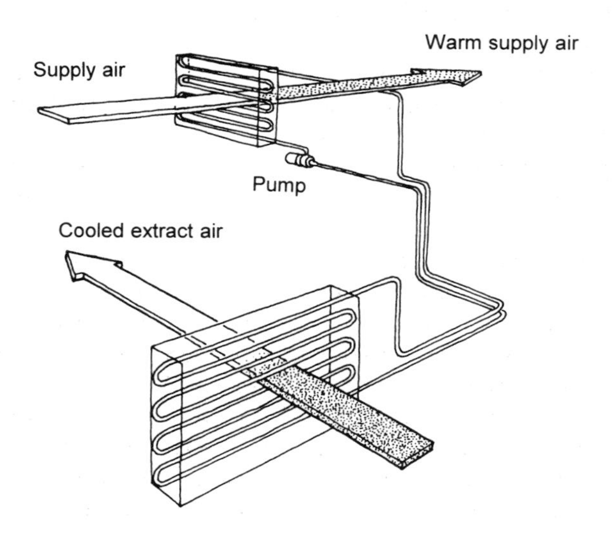

(c) Run-around coil

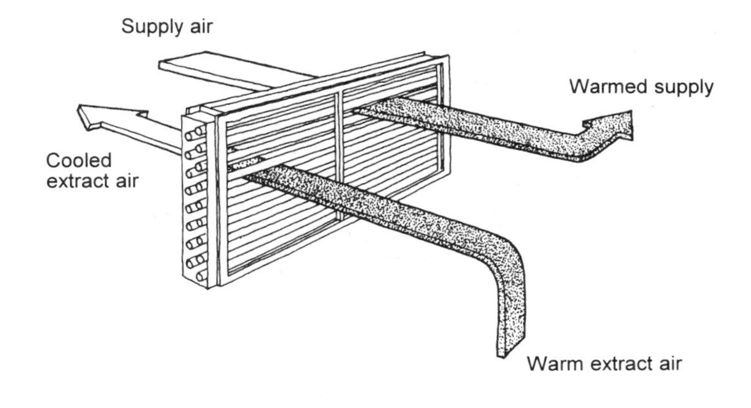

(d) Heat pipe heat exchanger

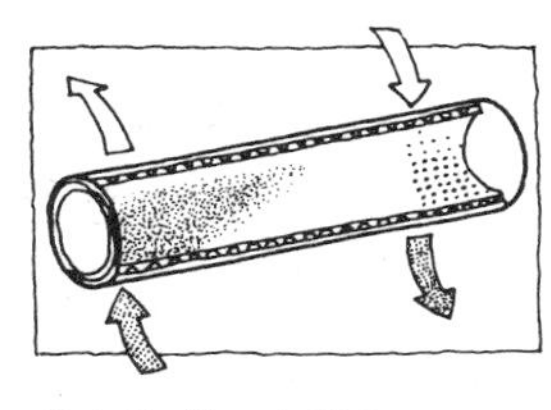

Detail of heat pipe

9.22 Heat recovery devices.

loss at the heat exchanger can be too great and the electrical energy requirement may outweigh the energy recovered (similarly, the CO_2 balance may be unfavourable).

Typical heat recovery devices are shown in Figure 9.22. In cross-flow heat exchangers one stream of air transfers its heat to another across a series of plates. Thermal wheels rotate in two streams of air; the medium of the wheel picks up heat in the hot stream and transfers it to the cool one. Run-around coils contain a liquid (often water) that is used to transfer heat. Heat pipes are sealed tubes containing refrigerant. If one end is placed in a hot area the refrigerant evaporates and takes the heat towards the cold end of the tube where it condenses; the refrigerant returns to the hot end by gravity and capillary action in the wick. Reference 33 provides a useful introduction to these and other methods of recovering heat. All have been developed for use in mechanical ventilation systems and all tend to have high resistances (in part to increase their efficiency) in an approximate range of 150–300 Pa for the high velocities in mechanical ventilation systems.[35] Resistances are, however, likely to be lower because of the lower velocities in assisted natural ventilation systems. (In a recent project by the authors for a lecture theatre, run-around coils with a pressure drop of about 30 Pa and fans with duties of 1.6 m^3/s at 50 Pa were incorporated.)

Approximate maximum temperature efficiencies (i.e. the difference between the temperature of extract from the space and the exhaust from the heat exchanger divided by the temperature difference between the supply air to the space and the incoming air to the exchanger) are 75% for cross-flow heat exchangers and heat pipes, 85% for thermal wheels and 65% for run-around coils.[36] System efficiencies are lowered by the extra energy required to force air through the heat recovery device. A fuller discussion on this topic can be found in Reference 37.

One of the problems with most natural and hybrid systems is that the outlet and inlet airstreams tend not to be in proximity, making it more difficult to transfer heat from one to the other. Run-around coils are less affected by this problem and it may be economically possible to incorporate coils with wide spacings and, thus, lower pressure drops. Velocities will also be lower in hybrid systems and although this, too, will lower the pressure drop it will also decrease the heat transfer efficiency.

Where the inlet and exhaust can be brought closer together, or where the heat recovered might serve another space, the other alternatives could be considered. There may be some scope for retractable systems – for example, heat pipes that are only used during the winter and in summer are pulled out of the stack.

A degree of filtration may also be possible where fans are incorporated. Filters present very high resistances to air flows, and although they can function readily in mechanical ventilation systems there is almost no scope for their use in natural ventilation systems. Typical low-efficiency flat panel filters in mechanical systems (where duct velocities tend to be about 5 or 6 m/s) have initial resistance varying from, say, 25 to 250 Pa. As we have seen, this is much higher than the driving force of the stack effect and so generally, with natural ventilation systems, what is outside is what you will have inside (at least at the beginning). However, there is scope for coarse, lower resistance filters being used in conjunction with less powerful fans in assisted natural ventilation systems.

A general point in looking at natural, assisted natural and hybrid ventilation systems is that it makes sense to use existing low-resistance routes such as floor and ceiling voids, corridors and staircases. The key concern is that this needs to work in conjunction with the smoke control strategy.

9.6 Mechanical ventilation systems

Considerations of air quality and noise or stringent temperature and relative humidity requirements may necessitate standard mechanical ventilation systems, and these are the subject of a vast literature. Suffice it to say here that a ductwork system can be used

- in connection with central air-handling units, which may be just a fan but are more likely to incorporate a heater battery and may also have a cooling coil
- in connection with localized sources of a filter and heating, cooling or both.

One aspect that concerns us is the limitation of mechanical ventilation to deal with heat gains. Fans that bring in external air in the summer cannot lower the temperature of that air (indeed, because some of the energy needed to run the fans is normally transferred to the airstream, they will raise the temperature slightly). Thus, if a space is at 29 °C because of occupants, computers, lights and so forth, bringing in air at 25 °C will, at best, bring the temperature of the space down to 25 °C. In practice it is more likely that the temperature would only be

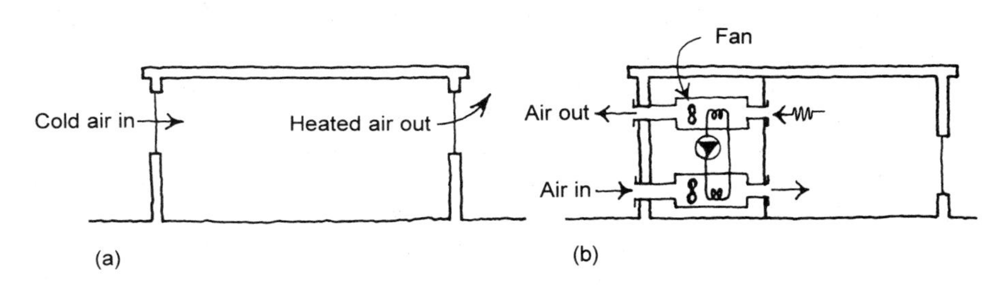

(a) Straight through natural ventilation – no heat recovery
(b) Mechanical ventilation with heat recovery

9.23 Ventilation models for the heating season.

brought down to 26 or 27 °C with the sizes of fans and air volumes likely to be used. And, of course, running the fans requires electricity and so results in CO_2 production (although in the nineteenth century, at least one bicycle-powered ventilation system was used in a London office building).

Mechanical ventilation with ductwork systems may be particularly advantageous in winter conditions when heat recovery is incorporated using devices such as those shown in Figure 9.22 and other equipment. Essentially there are two conceptual models, as shown in Figure 9.23. In model (a), which is the traditional type of building, air comes in and goes out via a variety of paths. The current tendency is to control these paths by more attention to construction details and by including such items as trickle ventilators. The great disadvantage of this model is that heated air is simply lost to the outside. In model (b) even more attention is given to control of the ventilation paths and great care is taken, for example, to seal the building, to incorporate high-performance windows which have very low leakage rates and to pressure test at completion to ensure that the construction is tight. It then becomes possible to introduce a controlled amount of air and extract a similar amount after recovering some of the heat it contains. This might seem to be an ideal system but it can be an expensive one. The additional capital cost of the fans, ducts and heat exchangers has meant that it has not been economically viable (given low energy costs). This is now under review but the outcome is by no means clear. Let us consider the simplest situation of a gas-fired heating system being used in both models (a) and (b). With the fans of the latter being run electrically (as they almost always are), the current cost of electricity is very roughly three to four times that of gas, and so effectively 'expensive' electricity (and a variety of equipment that must be paid for) allows one to recover 'heat' produced from cheap gas. The amount of heat that can be recovered depends on the installation. For each project an economic analysis needs to be made and this must include some assessment of the relative costs of gas and electricity.

A more recent concern has been total CO_2 production for these two models. As we have seen, electricity produces much more CO_2 than gas. With an inefficient heat recovery system the CO_2 production can actually be higher than in a straight-through ventilated building.

What is clearly needed is comparative research into the actual performance of some buildings of both types. One example of model (a) can be seen in De Montfort University's Queens Building (Chapter 13). Model (b) is represented by the new student residences at the University of East Anglia[38] and also by new

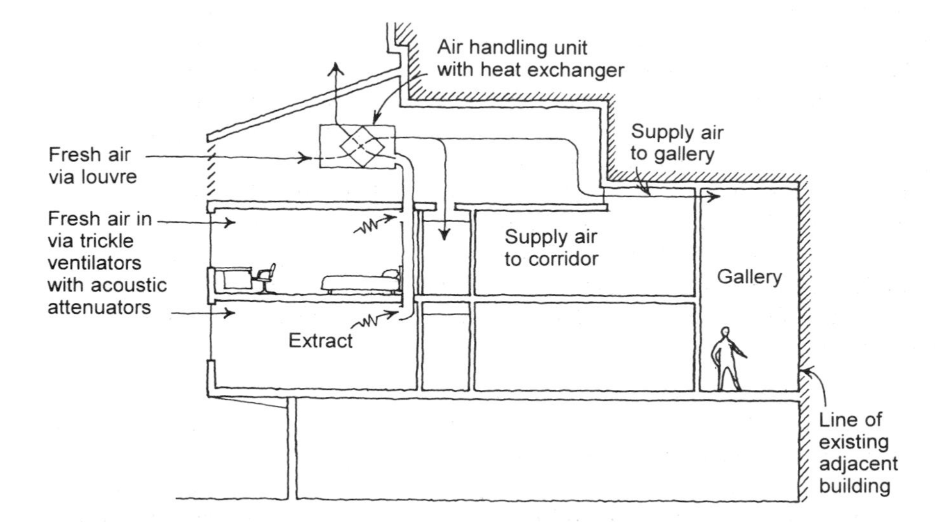

9.24 Ventilation with heat recovery at Christ's College Cambridge New Student Housing. (Architects: Architects Design Partnership.)

student housing incorporating an internal gallery at Christ's College, Cambridge. Figure 9.24 shows a simplified section through the building and a simplified schematic of the latter's ventilation system.

What factors tend to make it worth while to use a ducted air heating system with heat recovery? The first is, of course, availability of heat – a steady, ample supply of heat from people, equipment, lighting and, less reliably, solar gains is required. A fairly steady and significant demand is also, of course, required. A large number of people with the concomitant high fresh air requirement also helps because it means that the ventilation heat loss requirement will be significant compared to the fabric heat loss. Available space for the ductwork, is, of course, a prerequisite. And the more expensive the energy, the more worth while it is to recover it.

Recent developments in mechanical systems include displacement ventilation. Most mechanical (and natural) ventilation systems mix fresh air with indoor air to alter the temperature of the room air and reduce its level of pollutants. Usually this means that some of the pollutants will be breathed in. However, it is possible to direct air at people as, for example, in displacement ventilation where cool fresh air is introduced at low level and flows across the floor without a great deal of mixing. This pool of fresh air is pulled up by the convection currents generated by body heat (and computer and other low-level heat sources, if present). The polluted, warm air rises to high level above the occupied zone and is extracted. Obviously, the floor should be clean and the cool air should not be so cool as to make the occupants' legs cold.

9.7 Ventilation with cooling

If we now move to the extreme right of Figure 9.1, with the aim of keeping the internal temperature no higher (more or less) than ambient temperature, at some point it becomes worth while to consider the addition of a cooling coil to the mechanical ventilation system. If the aim is to keep the internal temperature below ambient temperature, a cooling coil (or some similar device) is almost obligatory. (The exception to this would be if the heat gains were low to moderate, ventilation requirements low and if the thermal mass of the building were high. The pyramids and King's College Chapel use no cooling.)

For the past 50 years or so where cooling has been provided, for example in many office buildings around the world, it has tended to be based on CFC and, more recently, HCFC refrigerants. As we have seen in Chapters 2 and 6, most of these lead to depletion of the ozone layer and so the more harmful CFCs are currently being phased out to be followed by a number of HCFCs. At present alternatives include:

1. Absorption cycle air-conditioning which uses, for example, ammonia as the refrigerant. This type of air-conditioning involves the absorption of the refrigerant vapour in another substance, hence its name. (The more common cycle is the vapour compression type that is used, for example, in most buildings and at home in refrigerators and freezers.) The absorption unit may have gas as its basic energy source, in which case it, of course, produces CO_2. This argues for a building design that avoids the need for cooling, if at all possible. More sophisticated and often large-scale systems can link combined heat and power plants to absorption cycle AC units.[39]
2. Ground water as a source of cooling (see Chapter 14 for a typical example and Reference 40 for others).
3. Vapour compression air-conditioning systems using the next generation of HFCs (which have zero ozone depletion potential as shown previously in Table 6.1, although they do have a significant global warming potential) or other substances. These systems are being introduced gradually and more will be learned about their performance in the next few years.

9.8 Controls

Controls are required to provide a comfortable environment with an economical use of resources – particularly energy. In the past this has been strongly influenced by costs and, with energy being cheap, relatively little was spent on controls. Now a combination of environmental problems and rising energy costs is forcing designers to give more consideration to controls.

There are a number of points that apply generally to controls. Firstly, the mechanical and electrical systems of a building and their associated controls should be as simple as possible, consistent with the need to meet requirements of efficiency, comfort and cost. A building that does more of the work will reduce the need for elaborate mechanical and electrical systems.

The more tightly something needs to be controlled, the more complicated, expensive and prone to breakdown the control system is likely to be. It is normally sufficient to let internal temperatures vary somewhat according to external conditions. This looser approach can minimize problems and save energy but it does require careful consultation among architect, engineer and client.

Try to imagine how the building will be used. Individuals want some control over their environments and need to be able to adjust them to avoid solar glare, or draughts, or uncomfortable temperatures. The needs of the individuals may conflict with those of the maintenance staff or with the operation of a central control system, and in most cases individuals should be able to override the automatic systems. In the future we are likely to see computer systems that allow such overrides but inform the individual if the action taken is not sensible.

Controls should be used to ensure that energy is produced only when it is needed and delivered only to the areas where it is needed. An important aspect of this is dividing the building into zones and ensuring that each zone is controlled suitably (but remember, too many zones are complex and costly). Zones can be thermal and so related to solar and internal heat gains, occupancy times, thermal response of the structure and type of heat emitter. Or they may be based on lighting, acoustic, smoke ventilation or other considerations.

Select control equipment that can be easily operated and locate it in an appropriate position. Controls must be comprehensible to those who are going to use them. As we become more computer-literate this poses less of a problem, but for the moment designers should speak with the people who will use the controls they specify to ensure that the controls match the people.

Controls should work with natural forces. One of the important developments in environmentally responsible buildings is to use controls to actuate lightweight elements that regulate internal and external forces. An example of this is a control system that operates lightweight dampers to let natural air flow through a space according to need (or demand). This 'intelligent' use of technology is an advance on the Jurassic forms of large air-handling units distributing significant volumes of air via extensive ductwork systems.

If we now look more specifically at controls for our services systems we can discuss them broadly by category.

Heating

Starting with heating, thermostatic radiator valves (TRVs), as shown in Figure 9.25, when fitted to heat emitters such as radiators or natural convectors will save energy provided, of course, that they are used. These valves work by restricting the flow of water to the emitter as the room temperature increases. An alternative that is somewhat more expensive is a thermostat in a space. This works in conjunction with an electrically operated valve (usually on–off) so that when the room temperature reaches the set point, flow to the heat emitters is shut off. Thermostats can be linked to time switches, and Figure 9.2 has shown a basic version of such an arrangement. Thermostats or temperature sensors linked to electrically operated valves can, of course, be used to control the entire range of heating emitters used in hydronic systems.

If we examine the operations of the boiler plant and heating circuits and move towards larger installations, energy can be saved by:

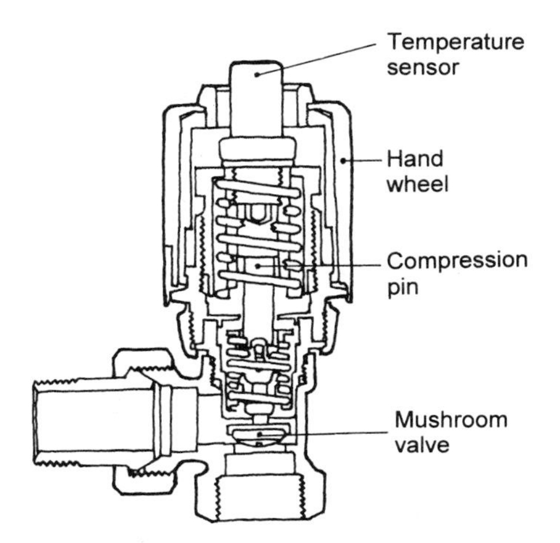

9.25 Thermostatic radiator valve.[41]

- including an optimizer in the system
- compensating the flow to each heating zone.

Optimizers switch the heating system on (or off) at a variable time that depends on outdoor and internal temperatures. They often incorporate 'learning' devices that can sense how the building responds. Thus, they are able to start the boilers earlier on a cold morning than on a warm one. This ensures that the building will be comfortable by the required time and that only the energy needed to achieve this is used.

Compensating the flow refers to lowering the flow temperature at higher external temperatures. This improves energy efficiency by reducing unnecessary heat losses from pipework.

For larger systems numerous additional control strategies are available but they are too specialized to be covered here.

Ventilation – natural and assisted natural

In their simplest and traditional form, natural ventilation controls consist of manual opening and closing of windows and in some cases, such as nineteenth-century schools, ventilating panels built into walls and roof cupolas with opening flaps. A more sophisticated ventilation system, based on the auditoria at De Montfort's Queens Building (shown schematically in Figure 9.26), requires more elaborate controls.

Following the air path and starting at the inlet point, if there is any possibility of rain getting in dampers should close (either partially or completely) on a rain signal. The inlet air velocity may need to be controlled if high wind pressures can result in discomfort or papers blowing in the space, and this can be effected by a wind sensor. By incorporating a noise sensor it is also possible to ensure that peak noise is reduced.

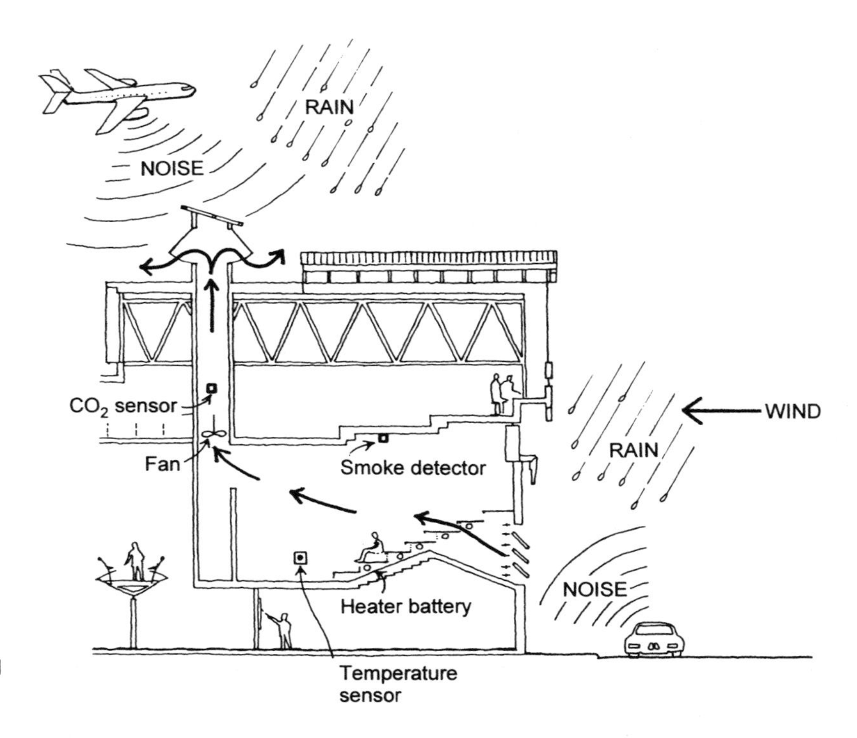

9.26 Automatic control strategies.

Incoming air needs to be tempered in the winter, and this can be done by controlling the flow to the heater battery by sensors close to the occupants. Once the air has entered the space it will affect the occupants and be affected by them. It is important to incorporate a high degree of user control and it is recommended that manual overrides be provided. To determine the amount of air that should flow through a space to provide adequate fresh air and avoid stuffiness, a CO_2 sensor can be used.

Temperature sensors in the space can determine whether the open area should be increased and whether a fan should be brought on to assist the natural ventilation.

Smoke or heat detectors may be used to control dampers and windows. Early contact with the fire officer is essential in developing a successful approach to ventilation and fire and smoke control. A common strategy, by which the space can be vacated quickly, is for all openings in the ventilation path to close automatically – they can then be left closed or opened when the fire brigade arrives and determines the best strategy.

The control systems may incorporate rain and wind sensors at roof levels. These can close openings as required to ensure that rain does not enter the building envelope and that excessive air movement in the space that might result from wind-driven ventilation is avoided.

An important benefit of such a control system is that it can incorporate night-time cooling, thus, the temperature of the space can be lowered to some optimum

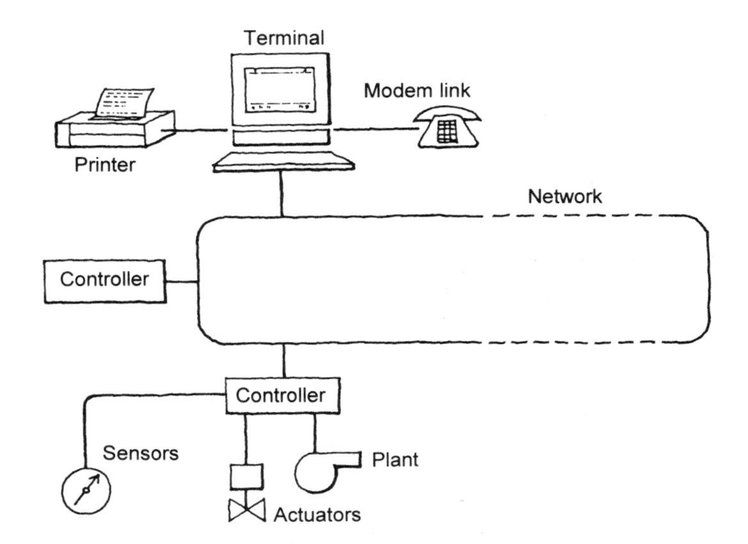

9.27 Typical configuration of a building management system.

value (related to its thermal mass and admittance) of, say, 15 or 16 °C by opening dampers and running the fan if necessary.

From the above discussion it is apparent that assisted natural ventilation systems like these are technically very sophisticated, even though they use much less traditional hardware such as ducts and air-handling units. With such systems natural ventilation can overcome one of its previous greatest disadvantages: lack of controllability.

Mechanical ventilation and air-conditioning controls

As with the systems themselves a vast specialized literature exists on their controls, and the interested reader should consult CIBSE and ASHRAE publications.[42,43]

Building management systems

Microcomputer-based building management control systems are key elements in the design of environmentally friendly buildings, especially larger ones. The principle of delivering energy only where and when it is needed assumes a network of sensors and a corresponding network of actuators such as switches and starters to run equipment or direct flows. Computers are, of course, well suited to this type of activity.

A typical layout for a building management system is shown in Figure 9.27. The system may vary in complexity but comprehensible simpler systems have obvious advantages in user-friendliness and cost. Systems whose computer routines are accessible and alterable by designers and users have definite attractions.

Guidelines

Heating

1. Specify high-efficiency and condensing boilers whenever possible.
2. Specify boilers with low NOx emissions.
3. The thermal response of the heating system must be matched to the building response and the pattern of occupancy.
4. Consider primary energy use and CO_2 emissions when selecting heating systems.

Ventilation

1. Use natural ventilation wherever possible, and avoid air-conditioning if you can.
2. Natural ventilation systems range from the simple to the complex. Do not underestimate the amount of thought required to make them work.
3. Assisted natural ventilation systems are likely to be appropriate, energy-efficient and reliable.
4. Give careful attention to noise and air quality.
5. Use heat recovery on extract air if you can prove it is economically viable and environmentally beneficial.

Ventilation with cooling

1. Cooling can be achieved without CFCs and HCFCs.

Controls

1. Design the building to do as much of the work as possible. Mechanical and electrical systems and controls follow afterwards.
2. For larger projects building management systems are often an intelligent solution.

References

1. Anon. (1990) *Guide for Installers of Condensing Boilers in Commercial Buildings*, Department of Energy, Efficiency Office Good Practice Guide 16, p. 2.
2. Lillywhite, M.S.T. and Trim, M.J.B. (1990). Gas-fired non-domestic condensing boilers. BRECSU Information Leaflet 18, BRE, Garston, p. 12.
3. Anon. (1993) *ASHRAE Handbook – Fundamentals*, ASHRAE, Atlanta, p. 15.11.
4. Anon. (1988) *Condensing Boilers in Local Authority Family Housing*, Department of Energy, Energy Efficiency Office Expanded Project Profile 284, p. 2.
5. See reference 1, p. 2.
6. See reference 1, p. 4.
7. Anon. (1989) *CIBSE Applications Manual: Condensing Boilers*, CIBSE, London.

8. See reference 3, p. 15.15.
9. Anon. (Undated; approximately 1993) *Dunphy Combustion Environmental Technology Guide*, Dunphy Combustion, Rochdale.
10. See reference 3, p. 15.15.
11. Baldwin R. *et al.* (1993) *BREEAM/Existing Offices: Version 4/93,* An environmental assessment for existing office buildings. BRE, Garston.
12. Stephen, R.K. (1988) Domestic mechanical ventilation: guidelines for designers and installers. BRE Information Paper 18/88. BRE, Garston.
13. Anon. (1993) *The Environmental and Economic Implications of All-Electric Houses. Summary of Conclusions*, Association for the Conservation of Energy, London.
14. John, R.W., Willis, S.T.P. and Salvidge, A.C. (1990) The BRE low-energy office: an assessment of electrical heating. BRE Information Paper 16/90. BRE, Garston.
15. Butler, D. and Howard, N. (1992) From the cradle to the grave. *Building Services*, **14**(11), 49–51.
16. Banham, R. (1969) *The Architecture of the Well-Tempered Environment*, The Architectural Press, London.
17. Richards, J.M. (1958) *The Functional Tradition in Early Industrial Buildings*, The Architectural Press, London.
18. Ibid.
19. United Distillers, Edinburgh.
20. Walker, L. (1981) *American Shelter*, Overlook Press, Woodstock, New York.
21. Lacy, R.E. (1977) Climate and building in Britain, BRE, Garston.
22. Anon. (1986) *CIBSE Guide A4: Air Infiltration and Natural Ventilation*, CIBSE, London.
23. Anon. (1978) Principles of natural ventilation. BRE Digest 210. BRE, Garston.
24. Ibid.
25. See reference 22, pp. A4–5.
26. Walker, R.R. and White, M.E. (1991) Single-sided natural ventilation – how deep an office. BRE. *Air movement and ventilation control within buildings. 12th AIVC Conference, Ottawa, Canada.*
27. Anon. (1994) Natural ventilation in non-domestic buildings. BRE Digest 399. BRE, Garston.
28. See reference 26.
29. Penz, F. (1990) *A Design Guide for Naturally Ventilated Courtrooms*, Cambridge Architectural Research Ltd, Cambridge.
30. Anon. (1988) *CIBSE Guide A5: Thermal Response*, CIBSE, London.
31. Baturin, V.V. (1972) *Fundamentals of Industrial Ventilation*, Pergamon, Oxford.
32. Anon. (1991) Code of practice for ventilation principles and designing for natural ventilation. BS 5925: 1991. British Standards Institution, London.
33. Hamilton, G. (1986) Selection of air-to-air heat recovery systems. BSRIA Technical Note TN11/86. BSRIA, Bracknell, Berkshire.
34. Anon. (Undated) *Heat Recovery with Heat Exchangers*, The Electricity Council, London.
35. See reference 33.
36. Anon. (1986) Heat recovery with heat exchangers. Electricity Council Publication EC3495/2.86. The Electricity Council, London.
37. See reference 33.
38. Evans, B. (1993) Airtight additions to the campus. *Architects' Journal*, **197**(17), 43–54.
39. Amberger, R.F. and De Frees, J.A. (1993) Retrofit cogeneration system increases refrigeration capacity. *ASHRAE Journal*, **35**(4), 24.
40. Evans, B. (1994) Cooling and heat from groundwater. *Architects' Journal*, **199**(5), 23–5.
41. Anon (undated) *Taco-Constanta Thermostatic Radiator Valve*, Tacotherm, Basingstoke.
42. Anon. (1985) *Automatic Controls: CIBSE Applications Manual*, CIBSE, London.
43. Anon. (1992) *ASHRAE Handbook: HVAC Systems and Equipment*, ASHRAE, Atlanta.

Further reading

Serive-Mattei, L., Babawale, Z. and Littler, J. (1993) Investigation of a domestic heating system with ventilation heat recovery: performance and integrity, in *Proceedings of the International Symposium on Energy Efficient Buildings, Leinfelden-Echterdingen, Germany.*

Spiers, K. (1992) Lessons in low energy. *Building Services*, **14**(6), 42–4.

Water, waste disposal and appliances

CHAPTER 10

10.1 Introduction

This chapter is a brief look at water, wastes and appliances. The one area examined in some detail is the provision of hot water services because of its importance in the design of larger, energy-efficient buildings.

10.2 Water

In the UK, availability of water is normally taken for granted and is only given any widespread attention in particularly dry summers with their resultant hose-pipe bans. But such an attitude is ultimately short sighted. Water is an essential natural resource whose collection, storage, treatment and distribution all have economic and energy costs. Schools alone spend approximately £70 million annually on water.[1]

There has thus been a slowly increasing acceptance of the need to conserve water by measures such as automatic flushing systems, self-closing and spray taps, and, of course, repairing leaking taps and leaks in the pipework. Recently, interest has turned to WCs with low water use. Current compact WCs use about 7-8 litres per flush but some commercially available units reduce this to 6 litres.

Less conventional approaches to conserving water include the collection and storage of rainwater. The Autarkic Housing Project at the University of Cambridge examined such systems in detail in the 1970s.[2] One approach is to use some of the rainwater directly for flushing and to purify another portion for drinking water by the use of filters or reverse osmosis units.

Recently, a house has been constructed in the UK that collects rainwater from the roof in copper gutters and stores it in polypropylene containers for bathing and other non-consumptive purposes.[3] About 5% of the rainwater stored is filtered so that it can be used for drinking and cooking purposes. Waste water is collected in a garden soakaway and used to irrigate the garden.

10.3 Hot water service

Hot water demands vary according to building type and use. Energy consumption is also governed by the building layout and the system. Typical annual figures for

Table 10.1 Approximate energy consumption and carbon dioxide emissions for hot water service systems[a]

Building type	*Primary energy use for hot water service provision as approximate percentage of total energy use (%)*	*Delivered energy use (kWh/m² yr)*		*Carbon dioxide production (kg/m² yr)*	
		Typical	*'Good'*	*Typical*	*'Good'*
Office	1–15	20	2–5	4	2
School[b]	5–25	35	5–10	7	1–4
House[c]	15–25	40–50	20	10–30	4–6

[a] This is for a variety of systems and is based on an extensive review which included References 4–10. The figures shown reflect varying contributions from gas and electricity.
[b] Based on an approximation of one pupil per 8 m^2.
[c] Based on four people in a 100 m^2 house.

energy consumption and CO_2 production for three types of building are given in Table 10.1.

Energy use and, consequently, carbon dioxide production are governed by a number of factors, including:

- water use
- system considerations such as
  - choice of fuel
  - concentration of demand
  - centralized vs decentralized systems
  - efficiency of water heating and control systems.

Water use will depend on the type of building, equipment installed and management techniques (to try to control use). A common starting point for analysis is to develop a demand histogram, as shown in Figure 10.1, which indicates the daily usage pattern.

Usage patterns can vary enormously: student residences have early morning and evening peaks while schools often experience a maximum demand at lunchtime for catering. Obviously the demand pattern can vary throughout the year, and again this is related to building type; schools will show a seasonal pattern but in office buildings consumption will be steady. Standard methods exist for estimating demand in a variety of building types,[12,13] and care should be taken to ensure that when applied they do not overestimate consumption for the building in question.

It is then important to evaluate the consequences of exceeding the likely demand; for instance, unavailability of hot water at home has very different consequences from unavailability in a hospital.

The most effective way of reducing energy consumption is reducing hot water consumption. On the domestic scale, a bath uses about 110 litres of hot water each time compared to a shower at about 40 litres. For larger buildings spray

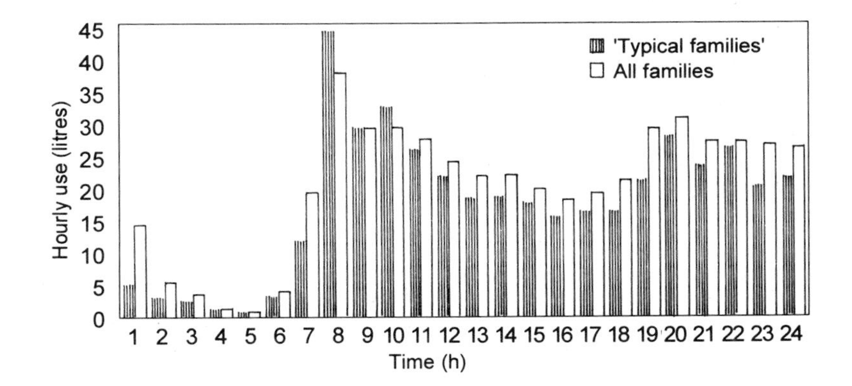

10.1 Demand histogram for US homes.[11]

taps and flow restricters can reduce demand. Equipment – for example, clothes washers and dishwashers – that uses less hot water is obviously advantageous as demand reduction normally has follow-on benefits of lower storage capacities, smaller boiler and pipe sizes, and reduced quantities of waste water.

Figure 10.2 shows an indicative decision tree for selecting from among the numerous systems available.

Choice of fuel

Presently in the UK the two main choices for fuel for hot water service systems are gas and electricity (see Chapter 7 for comments on other fuels). We have seen in Chapter 7 that, as a fuel, gas is both cheaper and has lower CO_2 emissions; however, the total system must be examined with the same precision as for heat recovery on extract air. Comparative studies for alternative systems in the same building are lacking, and studies of actual installations indicate that efficiency (defined as useful heat energy from the tap as a proportion of total energy delivered) ranges widely – figures for electric systems vary from 10–80% and for gas from 14–63%.[14,15] Normally one might expect properly selected point-of-use electricity systems to be more efficient than gas (in terms of conversion of delivered energy to useful heat) and to have lower capital costs but perhaps higher running costs and greater CO_2 emissions. Each situation, however, must be analysed in detail.

It should be noted that the choice of fuel may be related to site considerations. If gas supply and gas-fired boiler(s) are already included in the engineering services design, the economics of gas and electricity is different from a situation in which no space heating demand is being met by gas.

Concentration of demand

In laying out the building the hot water draw-off points should be brought together as much as possible. This will result in both simpler and shorter pipework

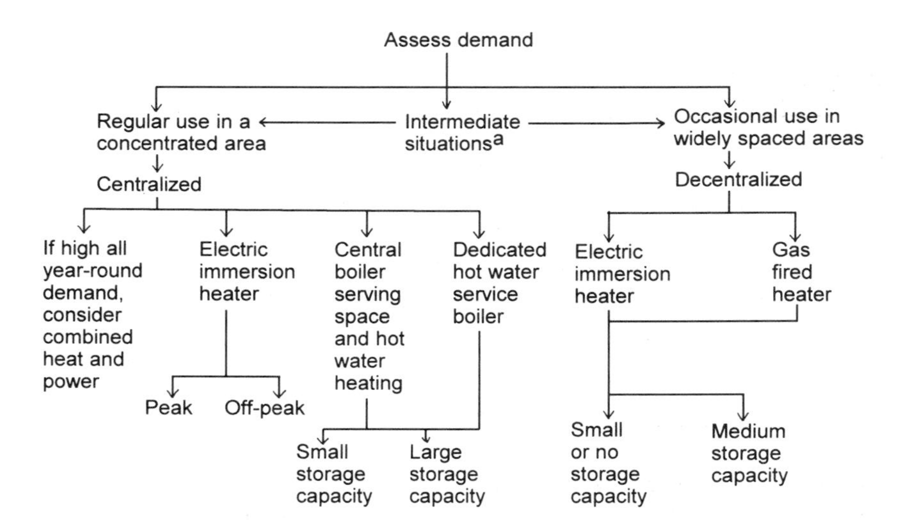

[a]Choice of systems for intermediate situations requires further details of each such situation.

10.2 Simplified selection of a domestic hot water system.

systems with less embodied energy, less pumping and reduced heat losses from pipework.

Centralized versus decentralized systems

At one end of the spectrum we have a concentrated area with a regular hot water demand served by a centralized hot water system. At the other end, we have widely spaced draw-off points with only occasional demands served by point-of-use heaters. If a centralized system were used in the second case, standing losses and perhaps circulating losses would be too great and capital costs could also be significantly higher. One element of the latter is that hot water pipework must be provided, whereas with point-of-use systems only a cold supply is required.

The point-of-use heater is likely to be electric if demand is small. If demand is greater, economics may favour a variety of gas-fired domestic water heaters. There are architectural considerations in decentralization. Gas heaters of course need a gas supply and flue, and many architects eschew having a proliferation of flues.

Maintenance is also a factor in the choice. Some clients prefer to have their services centralized because they find it makes maintenance simpler and less costly. However, this depends on the expertise of the maintenance staff, and where the staff are less technically orientated, very simple decentralized systems may be easier to deal with.

There is scope for combining these two types of system. For instance, if most of a building works normal hours, but a small area uses hot water during the night, then the area with the 24-hour water demand could be served by a local heater, allowing the system serving the remainder of the building to be turned off at night. This approach has proved successful in hospitals.

If a centralized system is selected, is it better to use one boiler for both space

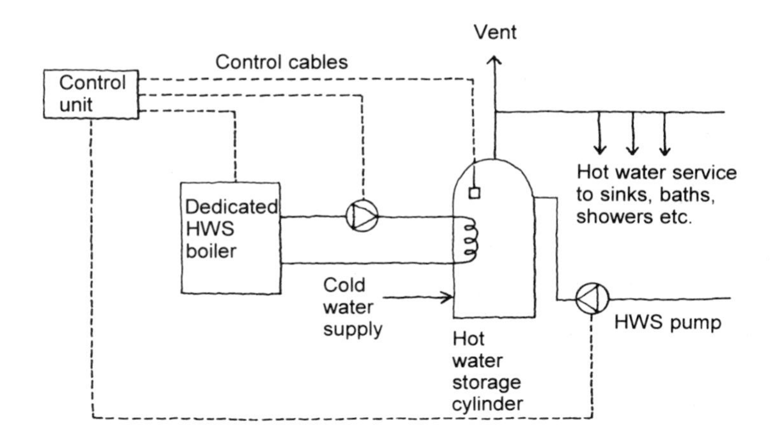

10.3 Hot water service system (simplified).

heating and the hot water service or to separate the functions? Because space heating is very seasonal and hot water service is much more constant, and because boiler efficiencies tend to fall on part load, there is a strong argument for a separate boiler for the hot water service. However, each case must be examined individually, and there is little reduction in performance on part load with some manufacturers' boilers.

Figure 10.3 shows a simple system with a dedicated hot water service boiler. In all such systems it is important to optimize the locations of plant and pipework.

The next consideration is likely to be boiler capacity vs storage capacity. Again, at one end of the spectrum we have a large boiler firing frequently and with no storage capacity and, at the other, a small boiler perhaps running almost continuously and with a large storage capacity, say, for 24 hours' demand. Most systems are, of course, between the two extremes and selection will depend on the demand profile and the space available for boiler and storage; to a much lesser extent it may depend on the peak fuel requirement and availability.

Boiler outputs should match the power required for the recovery time of the cylinders, and primary circuits should be pumped. This will allow reduction in the size of the pipes and thus will result in lower heat losses. Sequencing a series of boilers so that they operate near full load is common practice, but is often defeated by poorly configured controls.

Storage of water incurs a standing heat loss from the calorifiers and also from primary waterpipes, but can increase system efficiency by preventing the boiler from firing too frequently.

The temperature of the water in the calorifier and the distribution pipework shown in Figure 10.3 is an important issue. Although the higher the temperature the greater the standing losses, common practice is now to store water at or above 55 °C, but provision for mixing of water from bath or shower taps must be made to avoid scalding. One reason for a stored temperature of 55 °C is to discourage the growth of Legionella bacteria which can occur at temperatures between 20 and 46 °C (37 °C is the optimum for multiplication).[16] Another is to avoid the waste of water that results when large quantities of tepid water are run off before hot water reaches the tap. At the calorifier a circulating pump can be used to mix the water, thus preventing stratification and reducing the risk of Legionella developing.

At the pipework there are two options: one is to have a hot water service (HWS) pump which returns water to the calorifier, as shown in Figure 10.3; the other is to trace heat the supply pipework with an electric tape under the insulation surrounding the pipe. The pumped solution requires a pump, of course, and return pipework but tends to be less expensive to run and to consume less primary energy. Again, there are no easy answers and the alternatives require economic and environmental analysis.

Combined heat and power

Figure 10.2 shows combined heat and power (CHP) as a possible source of heating. If there is a significant demand for electricity and a concurrent demand for a hot water service, CHP can be ideal. An illustration of this is that CHP is proving most cost-effective in hospitals and small hotels. Note that 'alternative' sources of energy, which include CHP, may in fact compete with each other. At the De Montfort Queens Building (Chapter 13), initial studies looked at both CHP and active solar collectors to meet the hot water service load (and part of the space heating load), and CHP appeared more economically viable.

Controls

Boiler controls should bring the boiler on only when needed and water should not be kept at a higher temperature than required. A combination of time clock control and a control thermostat on the calorifier should be sufficient. The controls will also pick up the boiler and the HWS and calorifier pumps.

One of the most powerful aspects of controls is their ability to be adjusted to on-site conditions. This should be considered at every stage so that, for example, requirements can still be met efficiently if initial HWS estimates prove to be inaccurate in practice. Later, during use, being able to adjust the controls for the varying demand patterns that can occur will provide a more efficient installation.

Maintenance and management

An important but often overlooked aspect of efficiency is the maintenance and management of the installation. Those who manage the system should be familiar with how the designers intended it to function; they will then be in a better position to run it correctly initially and adapt it efficiently as conditions alter.

10.4 Waste disposal

Waste disposal systems that are not connected to mains sources include cesspools and septic tanks, and less conventional systems such as aerobic Clivus Multrum dry toilets and anaerobic methane digesters.[17] Cesspools are storage tanks which hold solids and liquids and need to be emptied regularly. Septic tanks retain the solid materials and discharge liquids to the surrounding soil, either directly or after biological filtration. Clivus Multrum dry toilets are ventilated waste disposal units that take WC wastes and produce, after a number of years, a compost that has been found to be pathogen free.[18]

Methane digesters are sealed devices which convert waste materials into an effluent sludge and produce methane gas in the course of the decomposition process.

The house referred to in section 10.2 above has a UK example of a working Clivus system, which decomposes the waste from a four-person family reducing it to a compost which is applied to the garden.[19]

10.5 Appliances

Appliances affect us in many ways, from choice of material to energy use. In Chapter 6 a CFC-free refrigerator was cited, which points the way towards careful selection of materials for appliances.

Considerable scope exists for energy conservation (and, where applicable, reduced water consumption) with improved appliances. In the USA, for example, residential refrigeration consumes about 7% of the nation's electricity![20] Energy consumption figures for appliances should be checked with manufacturers. Attempts are being made to introduce labelling schemes to encourage selection of higher efficiency equipment.

Guidelines

1. Conserve water by the use of efficient distributors such as automatic flushing systems and low water content WCs.
2. Select the hot water service system that is appropriate to the building and the demand.
3. Alternative waste disposal systems are available and may be appropriate.
4. Select appliances that use environmentally friendly materials and have low energy consumption.

References

1. Anon. (1993) *Managing School Facilities: Guide 1, Saving Water*, Department for Education. HMSO, London.
2. Littler, J.G.F. and Thomas, R.B. (1984). *Design with Energy*, Cambridge University Press, Cambridge.
3. Bunn, R. (1994) Living on auto. *Building Services*, **16**(7), 22–4.
4. Hampton, D. and Cooper, I. (1990) Saving energy in schools. BRECSU Information Leaflet 23. BRE, Garston.
5. Anon. (1992) Saving energy in schools: the headteachers' and governors' guide to

energy efficiency. BRECSU/EEO Best Practice Programme: Energy Consumption Guide 15. BRE, Garston.
6. Anon. (1989) Energy efficiency in offices: 100 Park Village East, London NW1. BRECSU/EEO Best Practice Programme: Good Practice Case Study 1. BRE, Garston.
7. Anon. (1990) Energy efficiency in offices: Cornbrook House, Manchester. BRECSU/EEO Best Practice Programme: Good Practice Case Study 17. BRE, Garston.
8. Anon. (1991) Energy efficiency in offices: South Staffordshire Water Companies, Wallsall. BRECSU/EEO Best Practice Programme: Good Practice Case Study 19. BRE, Garston.
9. Anon. (1993) Energy efficiency in offices: BRE low energy office. BRECSU/EEO Best Practice Programme: Good Practice Case Study 62. BRE, Garston.
10. Anon. (1992) Energy efficiency in new housing: Gifford Park. BRECSU/EEO Best Practice Programme: Good Practice Case Study 91. BRE, Garston.
11. Anon. (1991) *ASHRAE Handbook: HVAC Applications*, ASHRAE, Atlanta.
12. Ibid.
13. Anon. (1986) *CIBSE Guide B4: Water Service Systems*, CIBSE, London.
14. Skegg, V.E. (1982) Losses in DHWS distribution systems. *Health Service Estate*, **48**, 46–53.
15. Anon. (1988) Hot water service in offices. Electricity Council Leaflet EC4181/1-88. The Electricity Council, London.
16. Anon. (1991) CIBSE Technical Memorandum TM13:1991. Minimising the risk of Legionnaire's disease. CIBSE, London.
17. See reference 2, pp. 293–303.
18. Anon. (Undated) *Clivus Multrum: Is It Safe? Health Considerations*, Clivus Multrum USA, Inc., Cambridge, Mass.
19. See reference 3, pp. 22–4.
20. Meier, A.K. (1993) Field performance of residential refrigerators. *ASHRAE Journal*, **35**(8), 36–40.

Further reading

Leckie, J., Masters, G., Whitehouse, H. and Young, L. (1975) *Other Homes and Garbage*, Sierra Club, San Francisco.

Summary

This section briefly summarizes the preceding chapters and introduces the case studies that follow.

An overview

From the general to the particular and back has been the underlying approach so far. The general laws of physics govern your immediate environment and the chair you are sitting on is related to the global ecology.

Design starts with the site and continues through a choice of form, a selection of materials and development of the engineering systems that provide what the natural environment and building cannot.

There is no clear-cut distinction between architects and engineers in an environmentally responsive architecture. Frank Lloyd Wright said that the development of piped heating systems allowed him to articulate the building form. In current design, in part because of the development of improved insulation materials and the often high internal loadings of buildings, we are led to articulate the building to exploit the natural forces of Sun and wind. To do so requires a common understanding on the part of architects, engineers and builders. The case studies which follow are examples of a cooperative approach to the design and realization of a number of exceptional buildings.

It will be clear from them that there are no blueprints for low-energy, environmentally friendly buildings but that there are a number of ways of achieving comfortable conditions efficiently in both summer and winter.

Finally, it is worth while emphasizing that we are only starting – almost everything about an ecological approach to our environment remains to be done.

An introduction

The case studies selected illustrate a number of the principles discussed so far. Our debt to their architects and the other members of their design teams is enormous. The studies should be regarded as work in progress – within the short period they cover energy efficiency has improved and new techniques have been tested and proved successful. Among the important points they make is that an environmental approach not only need not constrain architectural creativity but can encourage it and that environmentally sensitive buildings can be of the highest architectural quality.

Part Two

RMC International Headquarters

CHAPTER 11

11.1 The site and the building

RMC House, completed in 1989, is the headquarters building for Ready Mix Concrete PLC, at Thorpe in Surrey. The site contained three existing houses and the architect's intention was to create a complex in which they were incorporated. Because the site is in the 'green belt' of London the architect strived to ensure that the headquarters were well integrated with the surrounds and that the public appearance was discreet. One approach was to design a largely single-storey structure with an extensive roof garden, as shown in Figure 11.1.

The building is about 10 000 m^2 in total, of which 4500 m^2 is offices. There are also training facilities in the form of various lecture rooms, an audiovisual studio, a laboratory, and an amenity complex with a swimming pool, two squash courts and a gym for the use of the staff. Finally, there is accommodation in the form of study bedrooms for the trainees.

11.1 Roof gardens.

One intention of the design was to keep the offices comfortably cool without air conditioning. This involved the application of passive techniques to reduce the heat gains in the offices and to mitigate their effects and was achieved by:

1. providing internal and external solar shading;
2. insulating the building to a high standard;
3. incorporating significant mass in the building for cool thermal storage;
4. use of the ground floor slab as a cooling element for the incoming air in conjunction with a mechanical ventilation system;
5. providing numerous openable windows to allow cross ventilation;
6. increasing the daylight penetration into the building to reduce the need for artificial lighting.

The roof build-up is 500 mm of soil on top of 100 mm of polystyrene insulation, on top of asphalt on a 150 mm concrete slab. This gives a U-value of 0.3 W/m^2 K which results in a low heat loss through the roof. The great thermal mass of the roof construction also provides a time lag of about 12 hours, which delays the external heat reaching the offices. This means that the heat of the day only reaches the offices at night, and that during the day the roof slab radiates, in a sense, the cool of the previous night into the offices.

The building is divided up by courtyards which allow daylight to enter the offices. One of these courtyards is paved with white tiles to increase the amount of daylight into the adjacent spaces. The walls that face the courtyards have full-height double-glazed patio doors; internal walls are mostly lightweight plasterboard, or glazing. Some of the deep plan internal areas have glazed rooflights, as shown in Figure 11.2; Figure 11.3 shows a similar area in plan.

In Figure 11.4 the perspective drawing shows the construction and the results of a calculation of the room admittance. (The concrete slab under the floor tiles contributes significantly to the room admittance.) Typical room admittances are 8 W/m^2 (floor area) K for lightweight constructions and 24 W/m^2 (floor area) K for very heavyweight construction.[1] External shading is provided by planter boxes

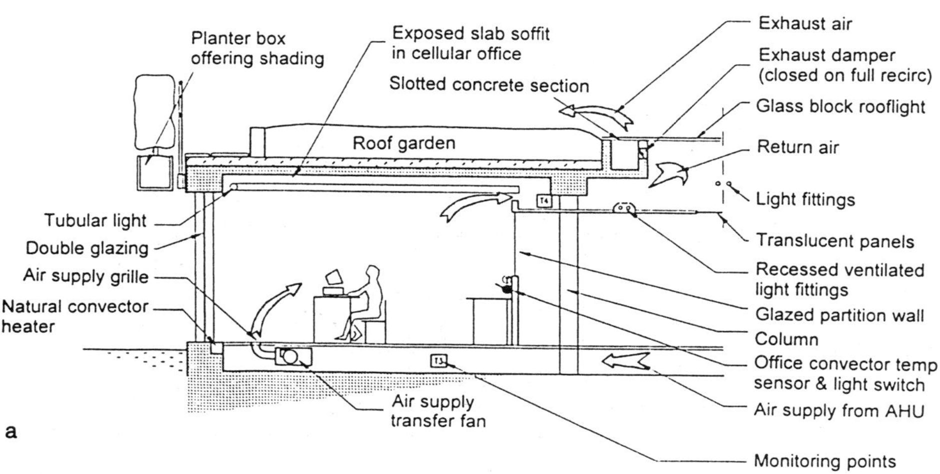

11.2 Section through a deep plan office area.

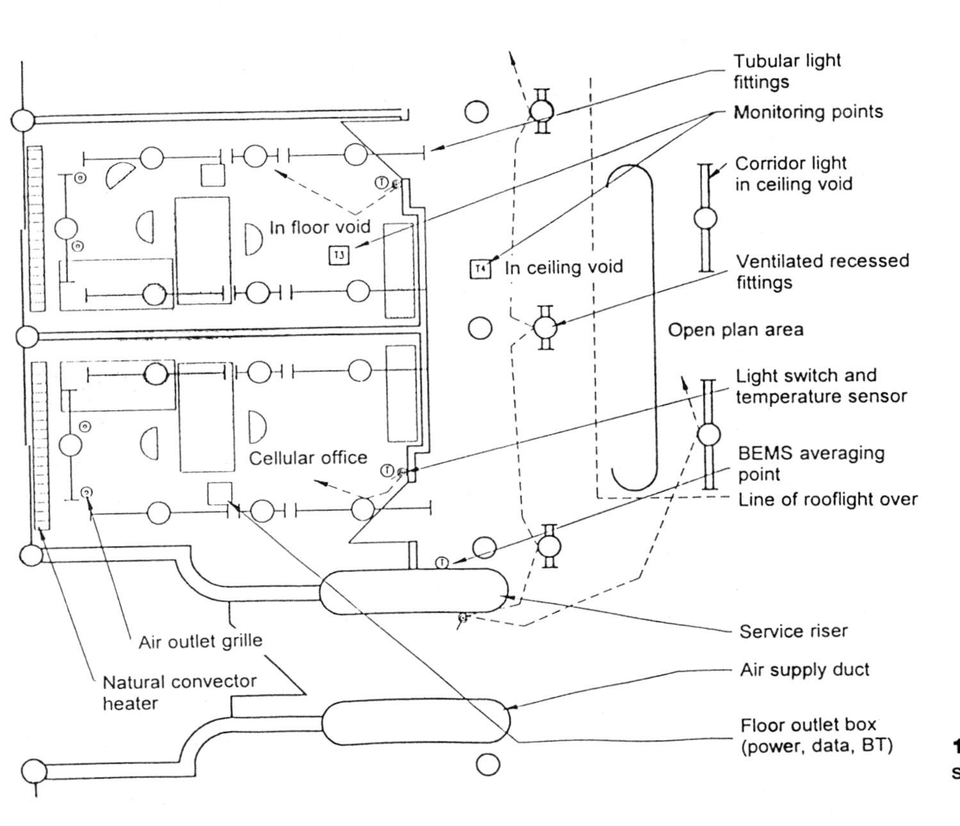

11.3 Office plan with services.

Medium weight room admittance 18 W/m^2 (floor area) K

Ceiling, concrete slab; internal walls, plasterboard; external walls, double glazing; floor, carpeted timber floor tile with concrete slab under

11.4 Typical office and room admittance. (Drawing by Edward Cullinan.)

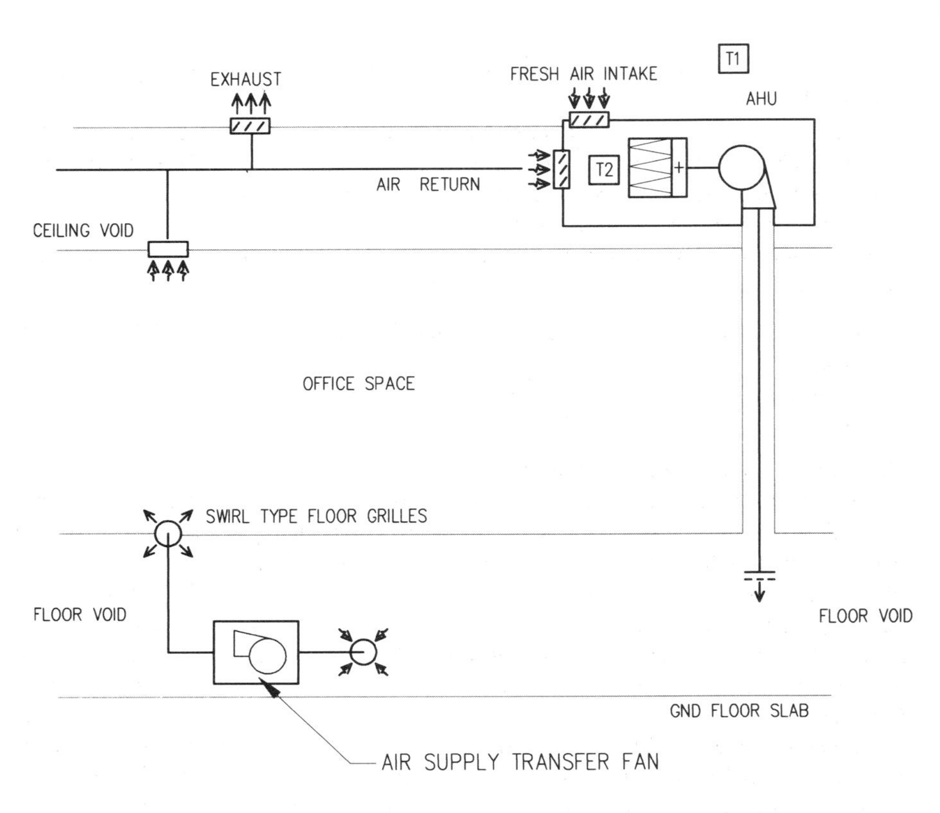

11.5 Schematic of mechanical ventilation systems.

which overhang the roof edge. Patio doors have internal venetian blinds. The core areas of the building have plasterboard false ceilings, but the perimeter offices have exposed concrete soffits, thereby allowing the heavyweight roof slab to help moderate the office temperatures. In some corridor areas there are rooflights of translucent panels (as indicated in Figure 11.3) to reduce the need for artificial lighting.

The perimeter offices have natural convector heaters under the patio doors which offset the heat losses through the glazing. Hot water is provided by central gas-fired boilers.

The offices are ventilated either by individuals opening their patio doors, or by mechanical air-handling units (Figures 11.2, 11.3 and 11.5). There are a number of compact air-handling plant rooms on the roof of the building, each of which serves roughly 500 m^2 of office. The air-handling units (AHUs) supply air down false columns to the floor void, which then acts as the supply ducting throughout the offices. The air is drawn from the void and blown out to the offices by small fans which serve four swirl type air outlet grilles. Air is extracted from the perimeter offices at high level in the false ceiling area, which is over the central offices. This area acts as a route for this extract air back to the rooftop plant rooms. From the central false ceiling area the air is either returned to the air-handling plant rooms to be recirculated, or is exhausted to atmosphere through specially designed motorized dampers under the rooflights. The air from the central office areas is extracted through the light fittings into the false ceiling zone. This removal at source ensures that little of the heat that is generated by the lights reaches the occupants.

The air handling units were sized to provide six air changes per hour to the offices or about 5 litres per m^2 floor area. This air flow was calculated to result in a maximum internal temperature of 25 °C when it reached 27 °C outside in conditions other than extended heatwaves.

The air-handling units generally recirculate the air, providing a small percentage of fresh air. The exceptions to this are:

- during the heating-up period in the morning when the air is 100% recirculated
- in summer when free cooling is possible, and in this case the air is 100% fresh.

Free cooling describes a situation when the offices are above their temperature set point and so require cooling, and the external air temperature is lower than the internal air so that cooling is possible. Cool air is taken from outside and passed over the ground floor slab under the raised floor. This is normally done at night by running the ventilation system and results in a process of cool thermal storage in the ground floor slab and the roof slab, which both have significant thermal mass. Perhaps the simplest way to think of what happens is to consider the slabs as storage coolers (rather than the storage heaters with which we are more familiar).

At night the fans will only run if there is a sufficient temperature difference between internal and external air to make their operation efficient and economic. If it is hotter outside than inside, the fans will not run as no cooling can be obtained. If it is only slightly cooler outside, then it is still not beneficial to run the fans as the heat gain to the air from the fan motors will heat the air to above the temperature of the internal air.

In some cases it is beneficial to encourage stratification of the temperatures in buildings in summer. Stratification is the situation whereby air at higher temperatures lies in layers above air at lower temperatures. Generally, but depending on the geometry of the space and ventilation rates, occupants will tend to be in the lower cooler areas. In this building this effect was intentionally reduced so that the roof slab could be kept as cool as possible in order for it to radiate to the occupants the following day. (In winter stratification is not desirable because the heat is needed at low level.)

The lighting in the cellular offices consists of suspended single-tube pipeline type fittings (Figure 11.2). The open plan offices have recessed, ventilated, twin-tube fluorescent fittings and bare single tube fittings between the translucent ceiling panels and the rooflights. The light level is about 400 lux in the central areas and about 600 lux in the cellular offices, with installed loads of 16 and 22 W/m^2, respectively. These figures are somewhat high because the lights provide some uplighting as well as downlighting, and, to a lesser extent, because high-frequency ballasts were not used.

The uplighting is less efficient because the reflectance of the ceiling is only about 60%, thus 40% of the uplighting energy input is lost immediately.

Lights in the cellular offices are individually switched. In the open plan areas

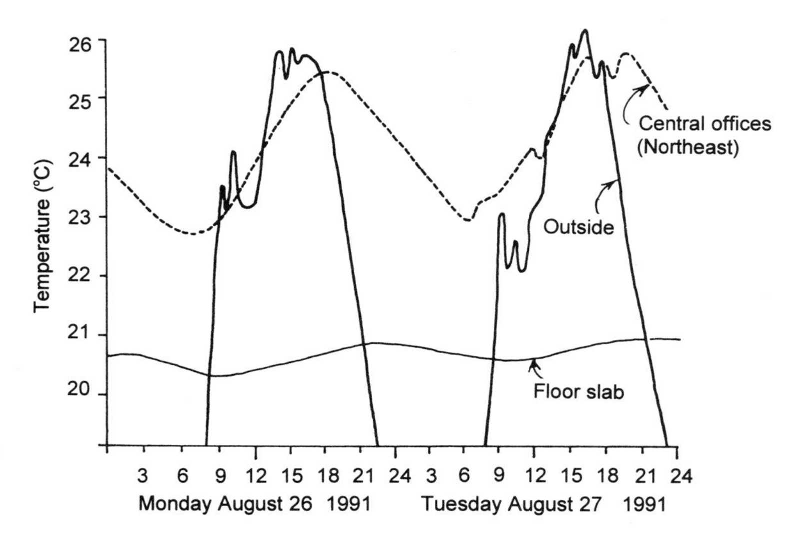

11.6 Temperature data for the central office area.

lights are switched centrally, but the switches are always local to the areas served so that the occupants are encouraged to switch off lights that are not required.

Most of the occupants of the cellular offices are senior executives, with administrative staff in the open plan areas. The occupancy is one desk per 20 m^2 in the cellular offices, and one desk per 30 m^2 (including circulation space) in the open plan areas giving a low loading from occupants of roughly 4 W/m^2.

Generally, in the building there are centralized photocopying facilities. Together with distributed small machinery such as personal computers and coffee machines, the loads are again low at about 7 W/m^2.

11.2 Internal conditions

The temperature inside the building is very stable, with an amplitude (maximum to minimum) of less than 3 °C on a hot day as can be seen in Figure 11.6. This stability is greater than might have been expected and is attributable to the large area of high admittance surfaces both directly connected to the space, such as the (coffered) ceilings, and indirectly connected via the ventilation system, such as the floor slab. The temperature of the floor slab remains very constant at 20–21 °C.

Other monitoring has shown that when the external air is at 28 °C, the air from the floor grilles is 2.5 to 3 °C below that temperature. The fans themselves raise the temperature of the air about 1 °C so the cooling effect of the slab is 3.5 to 4 °C. Given that the temperature difference between the external air and the slab is roughly 7 °C (from about 28 °C to 21 °C) the heat transfer efficiency is very good.

Analysis of extensive measured data has shown that the minimum ground floor slab temperature lags behind the minimum external air temperature by about 10 hours. Further examination also showed that the slab can take 4–6 days to

Table 11.1 Comparative delivered energy consumption data (kWh/m² yr)

Building	Heating and hot water services[a]	Fans, pumps and controls[b]	Lighting[b]	Equipment[b]	Other[b]	Total electrical
RMC offices	114.2	17.2	47.3	13.5	4.2	82.2
Typical office	200	21	53	16	5	95
Good practice office	95	13	35	16	4	68

[a]Provided by gas.
[b]Provided by electricity.

respond to a heat wave. Thus, it can be concluded that the ground slab functions effectively as a buffer or store.

11.3 Energy consumption

Figures provided by RMC show that the building used 1484 MWh/yr of electricity and 2676 MWh/yr of gas on average in 1991 and 1992. Table 11.1 gives a comparison of the energy consumed in the RMC office area and 'typical' and 'good practice' office data from the Energy Efficiency Office;[2] the RMC figures have been estimated using data from the building management system and details of the installed loads.

The figures in Table 11.1 indicate the generally good performance of the building. Electrical energy consumption is somewhat high because of the lighting as discussed previously.

11.4 Conclusions

The offices at RMC House are performing well and are reasonably energy efficient. In summer where the occupants have access to patio doors they have control over their own environment which leads to a high degree of occupant satisfaction. There have been some complaints from the people in the central area of the building, mostly due to a feeling of stagnant air, and probably exacerbated by the lack of outside awareness and individual control.

The high thermal mass of the building in conjunction with the underfloor ventilation system produces a stable floor slab temperature and comfortable conditions in the offices without the need for air conditioning.

There is no spare capacity, available with the flick of a switch, in the performance of a passively cooled structure. Therefore it is important to fine tune a number of aspects of the building. This is best done by monitoring performance and this is easiest with a building management system. The interest and expertise of the management and maintenance staff have also been key elements in this headquarters' success.

Project principals

Client	Ready Mix Concrete PLC
Architects	Edward Cullinan Architects
Services engineers	Max Fordham Associates
Structural engineers	Anthony Hunt Associates
Quantity surveyor	Raymond Hart & Partners
Landscape architects	Derek Lovejoy & Partners
Project managers	Pelmaks Project Management Services Ltd
Management contractors	Trafalgar House Construction Management
Mechanical contractors	Rosser & Russell
Electrical contractors	N.G. Bailey

References

1. Petherbridge, P., Milbank, N.O. and Harrington-Lynn, J. (1988) *Environmental Design Manual*, BRE, Garston.
2. Anon. (1991) Energy efficiency in offices. Department of Energy, Energy Efficiency Office Best Practice Programme: Energy Consumption Guide 19. HMSO, UK.

Further reading

Davey, P. (1989) Musique concrete. *Architectural Review*, **188**(1123), 59–67.

Grove Road Primary School

CHAPTER 12

The challenge at Grove Road Primary School, in the London Borough of Hounslow, was to minimize sound penetration to the classrooms of jets roaring over the school on their final approach to one of the runways at Heathrow (Figure 12.1). When the runway is in use, aeroplanes pass every few minutes only 150 to 300 m above the school.

Grove Road Primary School was the London Borough of Hounslow's oldest group of school buildings still in use. Its three Victorian buildings, originally built as separate schools for infants, girls and boys, imposed an undesirable constraint upon the requirements of modern teaching methods. The traditional high

12.1 Grove Road Primary School.

window sills restricted outside awareness, and the structural form and disposition of the buildings created lengthy internal circulation routes and negated integration between age groups. Two of the buildings were in a poor state of repair and all the buildings were surrounded by Tarmac without any soft areas.

The architect was appointed to rationalize the site, to bring the teaching spaces up to current standards and to introduce some landscaping for amenity and educational purposes. One of the overriding requirements of the brief was to ensure a very high level of sound insulation from the aircraft passing overhead.

The architect's proposal included refurbishing the existing building on the street side of the site to accommodate the school's administrative support and attaching the new classroom building to the rear. The floor level was raised by 600 mm to lower the sill height in the existing building. The other two old blocks were demolished and the site was cleared for a play area and landscaping. Valuable building materials such as slates and bricks were salvaged from the demolished buildings and reused.

This approach conserved the existing street elevation and restricted the new building element to that which was necessary to meet the new classroom needs.

To reduce heat loss and noise penetration, the new classrooms were designed as a compact, deep plan building formed as a semicircle round a central glazed courtyard, thus minimizing the surface area/volume ratio of the envelope. Advantage was taken of the single-storey plan by making full use of roof lighting to greatly reduce the requirement for artificial illumination.

To maintain the acoustic separation from aircraft noise, the roof, windows and walls were carefully detailed and constructed using standard building techniques. The roof construction was a slightly modified conventional design of roof tiles, two layers of heavyweight sarking felt, loft insulation acting as a sound absorbent, and double-layer plasterboard ceilings, all supported on wooden-trussed rafters.

By separating the two relatively light sound barriers (sarking felt and plaster-

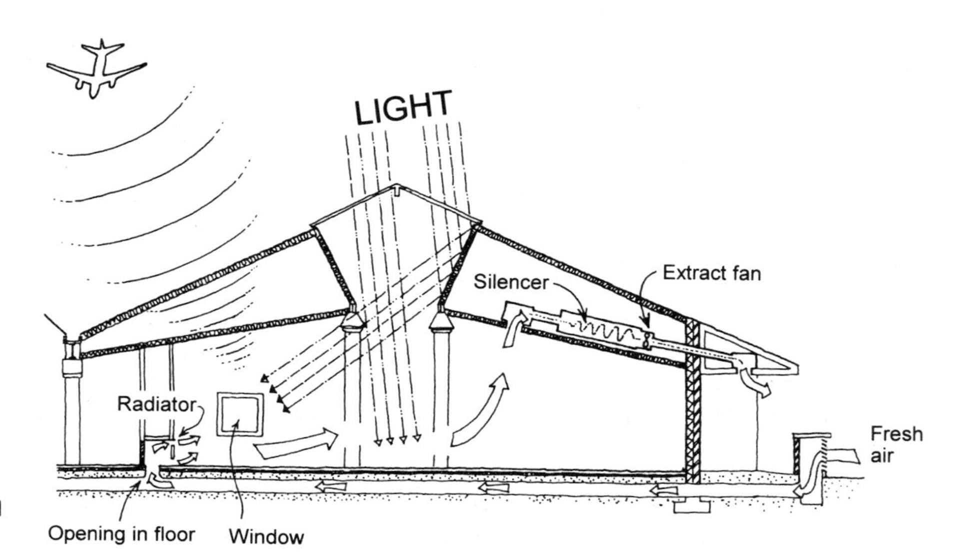

12.2 Environment and services at Grove Road Primary School.

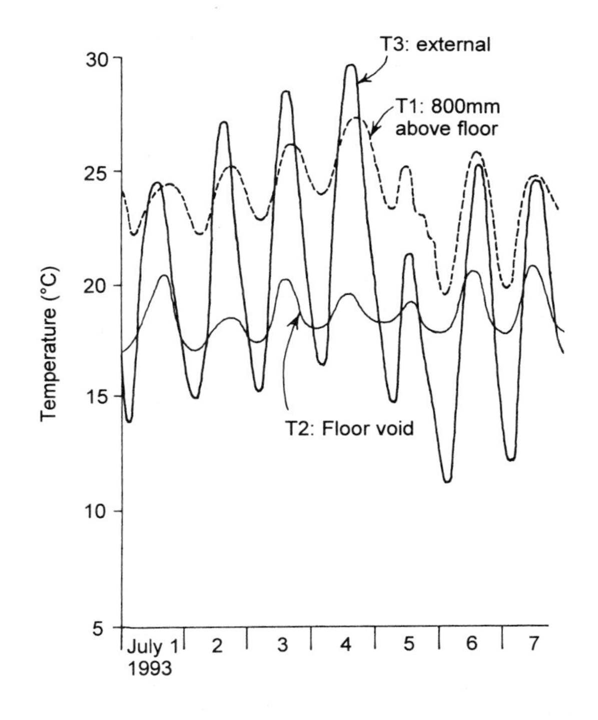

12.3 Thermal performance of Grove Road Primary School.

board ceiling) with a deep acoustic absorbent roof cavity, a high level of sound separation was achieved. The same principle was applied to the rooflights where two layers of glass were separated with a deep, absorbent void.

To maintain sound separation, the windows could not be opened, and another route had to be found for the entry of air. This was provided by the deep underfloor cavity; the void created by lifting the floor 600 mm was connected to the outside air by a number of ventilation paths around the perimeter of the building (Figure 12.2). As a result the cavity was freely ventilated with outside air. The fresh air from this void is drawn into the classroom through a hole in the floor behind a radiator. The tortuous air path from outside to the classroom is lined with unfinished building materials such as bricks and blocks which absorb sound. As a result, the aircraft noise is attenuated significantly by the time it gets to the opening in the floor. The overall result obtained by the building envelope was an attenuation of roughly 35–40 dB.

The quantity of fresh air can be controlled by the teacher in each classroom by adjusting a flap in the floor and switching on the extract fan. In the summer, the underfloor void acts as a thermal flywheel (a device to store energy and then release it) which precools the air entering the building. This complements the heavyweight internal masonry construction which minimizes summer overheating problems.

The thermal performance of the void and the space was monitored in July 1993. Figure 12.3 shows the external air, floor void and classroom temperatures; it can be seen that the floor void temperature is relatively stable compared to the outside and classroom temperatures and that the classroom temperature is below the peak outside temperature.

The central glazed courtyard is not heated but acts as a buffer space for the adjoining classrooms, providing light and solar heating when available. Air, pre-heated in the courtyard, is drawn into the conditioned spaces by means of a fan in a double-skin wall.

The fuel bills indicate annual delivered energy consumption figures of 150 and 18 kWh/m^2 for gas and electricity, respectively. These figures include the heating and the refurbished old building as well as the catering facilities. They compare with annual energy consumptions of 222 and 30 kWh/m^2 for gas and electricity used by the old school building before the refurbishment.

Grove Road Primary School significantly enhances the local environment with the minimum of building works and has peaceful, internal teaching spaces that are isolated from the disruptive effects of aircraft. The form and planning of the building also makes it very energy efficient. The building acts as a significant model for any building type, showing how a deep plan can work and demonstrates a low technology solution to noise pollution.

Project principals

Client	London Borough of Hounslow
Architects	Plincke Leaman & Browning
Services engineers	Max Fordham & Partners
Structural engineers	Jampel Davison & Bell
Quantity surveyors	John Smith Associates
Main contractors	Croudace Construction
Mechanical and electrical subcontractors	Lorne Stewart PLC

Further reading

Ostler, T. (1993) Passing the ecology test. *Architect Builder Contractor and Developer*, September, 10–11.

Queens Building, De Montfort University

CHAPTER 13

The Queens Building at De Montfort University, shown in Figure 13.1, which houses the school of engineering and manufacture, is one of the most exciting and innovative buildings to have been built in Europe in the last few years. The building represents a major shift from the tradition of university buildings that has grown up over the past 20 to 30 years. The design strategy chosen has been to look to the solutions of the past then, applying modern technology, to refine these solutions and make them the way to the future. In particular, the building has been designed to function without the use of air conditioning in spite of high internal heat gains in lecture theatres and computer laboratories. The result at De Montfort is a building that is wholly modern with a strong bias towards low-energy design which, with its red brick construction, fits well into the traditional Leicester cityscape.

13.1 External view of the building.

Table 13.1 U-values[a]

Element	*Construction*		*Thermal resistance (m^2 K/W)*	*U-value (W/m^2 K)*
External wall	External surface resistance		0.06	
	100 mm brickwork	(1.18 m K/W x 0.1 m)	0.118	
	100 mm Rockwool	(28.57 m K/W x 0.1 m)	2.857	
	190 mm concrete block	(0.61 m K/W x 0.19 m)	0.116	
	Internal surface resistance		0.12	
	Total resistance		3.27	0.30
Roof	External surface resistance		0.04	
	20 mm roof tile	(1.19 m K/W x 0.02 m)	0.024	
	32 mm cavity	(0.179 m K/W x 0.032 m)	0.006	
	5 mm felt	(2.0 m K/W x 0.005 m)	0.01	
	150 mm Rockwool	(28.57 m K/W x 0.15 m)	4.286	
	50 mm cavity	(0.179 m K/W x 0.05 m)	0.009	
	18 mm timber	(7.14 m K/W x 0.018 m)	0.128	
	Internal surface resistance		0.12	
	Total resistance		4.623	0.22
Internal wall	Internal surface resistance		0.12	
	100 mm concrete block	(0.61 m K/W x 0.1 m)	0.061	
	Internal surface resistance		0.12	
	Total resistance		0.301	3.32
Ground floor				0.19
Double glazing				3.4

[a]See Appendix A for the calculation procedure.

13.1 Building description

Construction

The building is, in general, four storeys high and of traditional red brick with matching mortar. The brick forms the outer skin of a cavity wall construction consisting of a 100 mm insulated cavity and an internal skin of dense fair face blockwork or calcium silicate brickwork. Floors are, in general, of precast double-tee concrete beams left exposed and painted. The roof structure consists of tiles on a timber subframe with either timber tongued and grooved boards or plaster-board internal soffits. Table 13.1 lists the U-values for the different elements.

Function

The building has three main areas, as shown in Figure 13.2.

The first area is the electrical laboratories at the east end of the building which make up the two four-storey 'fingers' that enclose the main entrance

Table 13.2 Maximum design internal heat gains

Space	Area (m^2)	Solar gain (W)	Equipment gain (W)		Ocupancy (100 W/person)	Total (W)	Total (W/m^2)
Electrical laboratories	170	4 500	Computers Lighting	6 000 1 610	3 000	15 110	88
Central building Auditoria	150	–	Audio-visual Lighting	500 2 100	15 000	17 600	117
Mechanical laboratory Machine hall	644	33 000	Equipment Lighting	10 000 25 000	2 500	70 500	109

courtyard. These house the computer laboratories of the electrical engineering school, which have high internal heat gains. Table 13.2 lists the design heat gains used in the calculations to define the necessary opening areas required to allow these spaces to be naturally ventilated. The floors have a shallow plan of approximately 6 m, which meant that as long as sufficient opening area could be provided they could be easily cross ventilated.

The second area is referred to as the central building. This is a deep plan, highly massed space containing, on the south side, classrooms at ground-floor level, a large double-height combined electrical and mechanical laboratory at first-floor level and staffrooms on the second-floor mezzanine and third floor.

The north side consists of two 70-seat lecture theatres below the double-height 150-seat auditoria at first-floor level. At third-floor level the drawing/design studios are traditionally designed to have a combination of top and north light.

The two sides of the central building are separated by a long narrow central concourse space which rises through the full height of the building and is top lit. This acts as the central communication route at ground- and first-floor level between the electrical and mechanical laboratories. In addition, it is bridged at third-floor level by flying walkways which link the staff offices to the drawing studios. The drawing studios overlook and are open to the main concourse volume.

The mechanical laboratories are the third main area and are located at the west end of the building. The main space is a double-height machine hall with a control room on a mezzanine level at each end. Adjacent to the machine hall at ground- and first-floor levels are smaller rooms which are dedicated to specific mechanical engineering functions such as welding, printed circuit board production and metrology studies. Owing to the nature of the work carried out in these ancillary rooms and the presence of fume cupboards to handle chemicals, two small air-handling units have been installed to provide these rooms with mechanical ventilation (but not cooling).

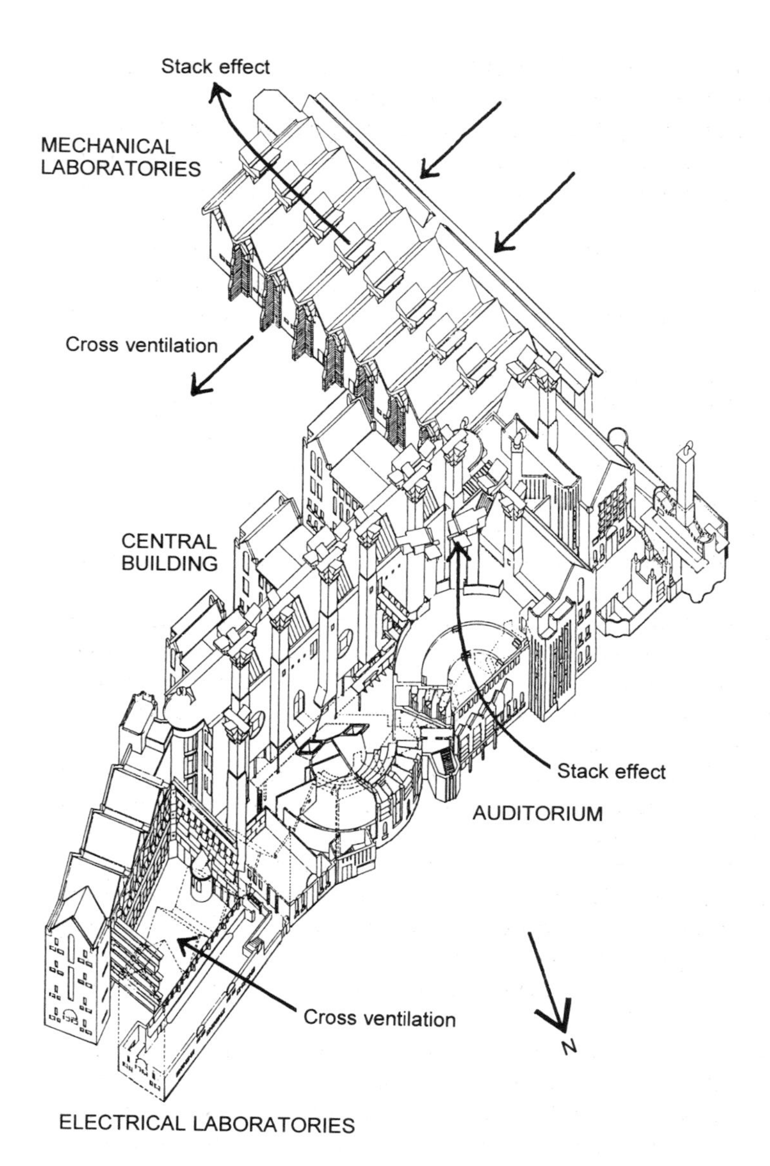

13.2 Axonometric of Queens Building.

13.2 Natural ventilation

One of the main design principles of the building was that it should be naturally ventilated because of the client's and design team's commitments to environmental issues.

The two auditoria proved to be the most heavily loaded areas from the point of view of internal heat gains, followed closely by the electrical engineering computer laboratories. Internal gains in the mechanical laboratories, and in particular, in the machine hall were more difficult to predict because although there are some very large items of equipment their use tends to be intermittent.

As we have seen previously (Chapter 9), in naturally ventilated buildings air moves because of pressure differences arising from stack effect and wind effects. Stack effect or buoyancy forces are caused by warm air rising and being replaced by cold air at low level. Wind effects arise due to wind flow around the building causing pressure gradients over the building envelope. Normally, both effects are in operation, but the contribution of each varies with the geometry of the space, the size and positions of air intakes and exhausts and the temperature differential between inside and outside.

The new engineering school has a range of different combinations of stack and cross ventilation. The electrical laboratories, because of their narrow plan, are cross ventilated and wholly manually controlled via opening windows. The window openings were sized using a low design wind speed of 0.5 m/s, which is the wind speed expected to be exceeded for more than 90% of the time. Table 13.3 shows the equations used in the calculations of the opening areas required for a typical cross-ventilated electrical laboratory using the heat gain figures given in Table 13.2.

The more deep plan spaces of the central building that are open to outside air on one face only are stack effect ventilated, using, in the case of the ground-floor classrooms and first-floor central laboratory, a combination of manual air intakes (opening windows) on the perimeter and automated air exhausts via stacks or openable rooflights deeper into the plan.

The auditoria which, as we saw in the worked example above, are also stack effect ventilated gave rise to the dramatic stacks above the roof line. The brief from the client for the auditoria called for the seating of 150 people in a space that could be blacked out. The thermal and acoustic environments did not have to meet the stringent criteria of an air-conditioned auditorium. This is one of the central issues relating to providing low-energy and naturally ventilated buildings – the client must be prepared to accept a more natural environment with its associated changes in temperature in the summer as the day warms up and, indeed, changes in noise levels as traffic passes by outside.

The occupancy, together with solar gains and gains from lighting and equipment, made the auditoria a main focus of the design. It was agreed at a very early stage that the lighting would have to be fluorescent to keep the heat gains as low as possible compared with more traditional tungsten sources. This decision also means lower maintenance costs as the lamp life of the low-energy fluorescent fittings used is approximately 8000 hours compared to 1000 hours for a tungsten source. The architects included the auditoria in their initial computer analysis of stack effect ventilation. These preliminary studies suggested that a loading of 100 W/m^2, an air inlet area of 2.5 m^2 and a stack height of 7 m would result in a peak internal temperature of 27 °C, given an external peak of 24.5 °C.[1]

Subsequent theoretical testing included salt water modelling by Cambridge Architectural Research Limited in conjunction with the University of Cambridge Department of Applied Mathematics and Theoretical Physics. A perspex model of the auditorium was built and immersed upside down in a bath of water. Salt

Table 13.3 Worked examples for calculating the opening areas required for natural ventilation using equations from Chapter 9

Stack effect – Path 3 (Figure 9.15) as applied to an auditorium

Under conditions where there is a 5 °C temperature difference between inside and outside and the height of the stack is 12 m

$$\Delta p = 0.043\, h\, (t_{int} - t_{ext})\text{Pa}$$

$$\Delta p = 0.043\, (12)\, (5) = 2.58 \text{ Pa}$$

$$A = \frac{A_i\, A_o}{(A_i^2 + A_o^2)^{0.5}}$$

$A_i = 4.9$ m^2, i.e. the free area of the air intake from the road; $A_o = 5.2$ m^2, i.e. the free area of the air exhausts.

$$A = \frac{(4.9)\quad(5.2)}{(24.0 + 27.0)^{0.5}} = \frac{25.48}{7.14} = 3.6 \text{ m}^2$$

$$\begin{aligned} Q &= 0.827A(\Delta p)^{0.5} \\ &= 0.827(3.6)(2.58)^{0.5} \\ &= (0.827)(3.6)(1.6) \\ &= 4.8 \text{ m}^3\text{/s} = 17\,280 \text{ m}^3\text{/h} \end{aligned}$$

The floor area of an auditorium is approximately 150 m^2 and the average height is 5.1 m, thus the volume is 765 m^3. This gives an air change rate of 23 per hour when there is a 5 °C temperature differential between inside and outside.

The fresh air requirement for 150 people in the auditorium is 10 l/s person x 150 = 1.5 m^3/s. This is significantly less than the air movement caused by the stack effect in this example, and in winter conditions the air intakes and exhausts can be opened only slightly to allow the ingress of enough fresh air for the people in the space. In winter the temperature difference between inside and outside is much greater than the 5 °C used in this example, and if the air intakes and exhausts were fully opened there would be a very high air flow through the space and a huge heat loss.

Wind effect – Path 2 (Figure 9.15) as applied to the top floor of the electrical laboratory

Cross ventilation by the wind alone is given by

$$Q = 0.61 A_w u (\Delta C_p)^{0.5}$$

where 0.61 is the discharge coefficient and A_w is the equivalent area (m^2) for wind-driven ventilation given by

$$\frac{1}{A^2_w} = \frac{1}{A^2_1} + \frac{1}{A^2_2}$$

In this case, assume ΔC_p to be 0.5 (ΔC_p is the change in the pressure coefficient; ΔC_p ranges from 0.1 for a sheltered site to 1.0 for an exposed site). u is the velocity, taken as 4.0 m/s, which is our assumed wind speed at a height of 10 m exceeded for 50% of the time (Appendix A).

$$A_1 = 2.535 \text{ m}^2 \text{ and } A_2 = 2.535 \text{ m}^2$$

$$\frac{1}{A^2_w} = \frac{1}{(2.535)^2} + \frac{1}{(2.535)^2} = 0.31$$

Table 13.3 (continued)

$\therefore A^2_w = 3.22$ and $A_w = 1.79\ m^2$

$Q = (0.61)(1.79)(4.0)(0.5)^{0.5} = 3.10\ m^3/s.$

Volume of electrical laboratory

$= 170 \times 3.3$

$= 561\ m^3$

$3.10\ m^3/s \times 3600 = (11\,163\ m^3\ \text{of air/h})/561$

$= 20$ air changes/h.

To estimate the temperature rise in the space at the flow rate available, and taking the total heat gain in the electrical laboratory from Table 13.2 as 15.11 kW or 15.11 kJ/s,

$$\frac{15.11}{1.2 \times \Delta T} = 3.10\ m^3/s$$

(where 1.2 is the specific heat capacity of air measured in kJ/m^3K).

Thus the temperature rise in the space ΔT is 4.1 °C. If the amount of air flow calculated in this way gives a temperature rise in the space which is considered unacceptable, then further openings should be introduced to increase the air flow which will, in turn, reduce the temperature rise. In practice, both the heat loads from computers and the temperature rises have been less.

solution was then dyed and injected into the perspex model and the resultant water flow patterns were filmed and analysed. The density of the salt water could be increased or decreased to represent higher or lower heat gains, respectively. The filmed results of the tests were inverted so that the salt water dropping could be seen as warm air rising. The model showed that connecting more than one space into a stack could result in air flow from one room to the next rather than upwards to exhaust at the top of a stack. On this basis individual spaces were given dedicated stacks. This, in turn, coincided with the requirements of the local fire prevention authority for smoke control in a fire situation. The weakness of the perspex model is that it could not take account of admittance (i.e. the use of the building fabric as a thermal store).

In addition to the salt water model, the School of the Built Environment at De Montfort University carried out computer simulations using the ESP package. The authors carried out the standard Chartered Institute of Building Services Engineers admittance procedure and stack effect calculations. Table 13.3 shows simplified versions of the equations used for the stack effect calculations.

These calculations were done both as a check on the more sophisticated techniques mentioned above and because design liability lies with the design team. A figure of 10 litres of fresh air per second per person was used for odour control and we chose an external design day of 25 °C maximum, 19 °C average and 13 °C minimum. The client was informed and agreed with this choice of day. The results showed that, in general, peak internal temperatures would not rise by more than 4–5 °C above peak external. This was a conservative figure based on temperature difference only (i.e. stack effect calculations) and not wind effect. As a back-up for the occasional very still, hot day a punkah fan of duty 1.7 m^3/s at

Table 13.4 Results of natural ventilation analyses for the auditoria

Technique	*Peak internal temperature*[a] *(°C)*	*Comment*
Saltwater modelling	29.8[2]	Stack effect only, thermal inertia not accounted for
Computer modelling	28.9[3]	Wind effects were simulated in addition to stack effect
Admittance procedure	29.6	Stack effect only; the technique is relatively crude and ventilation rates are rough estimates

[a]Temperatures have been adjusted relative to an external peak temperature of 25 °C (based on a day with a 13 °C minimum and 19 °C mean).

10 Pa was included in one of the two stacks in each of the two auditoria to provide some minimal mechanical assistance.

A comparison of the results from the various calculation techniques is shown in Table 13.4. Actual results, as shown in Figure 13.8, are in fact significantly better than predicted.

Figure 13.3 shows a section through one of the auditoria. Air enters at the street side through large openings protected from the weather, passes through motorized volume control dampers at the building envelope line and through an

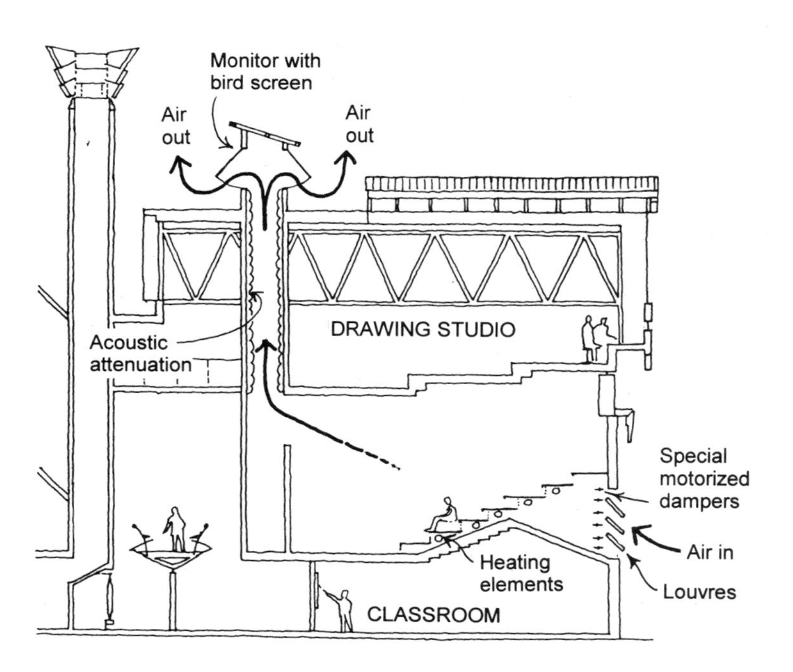

13.3 Air passage through auditorium.

13.4 Stacks at roof level.

acoustically lined plenum and is distributed through the void under the seating. It passes over finned heating tubes suspended under the seats, out under the seats and then through a grille made of aluminium mesh. There are no air filters, which would create too high a resistance to air flow.

The air that is heated by the occupants and other internal gains, such as from lighting and audiovisual equipment, then rises and travels into the diamond-shaped exhaust stacks which also serve as supports for the drawing studio roof trusses. The total cross-sectional free area of the two stacks is 4 m^2. At the top of the stacks, as shown in Figure 13.4, are automatically controlled opening windows with a total free area of approximately 7.9 m^2. The opening area is roughly based on a rule of thumb for chimney tops that the area of the vertical opening faces should be at least twice the flue cross-sectional area. The discharge coefficient for windows varies from approximately 0.57 at an opening of 60° to 0.63 at 90°. There is a slight overhang at the top of the stacks to help reduce the chance of rain entering.

The key points in the design of the auditoria are:

1. The spaces have high thermal mass and high ceilings. In summer the building structure is precooled at night by allowing air movement through the room. This helps to lower the day-time temperatures.
2. The air path from the air intakes, through the space and out through the stacks, is of low resistance, i.e. the pathway is mainly unobstructed.
3. The stacks terminate approximately 3 m above the roof line to avoid local turbulence. The air exhausts via automatic opening windows at the stack tops.

13.3 Lighting

Daylighting

The building has an extended perimeter to produce good daylighting for all spaces. In the early design days the architects tested a 1:50 scale model of the building under the artificial sky in the School of the Built Environment at De Montfort University and, to a lesser extent, tested a smaller scale model on a heliodon. In addition, daylight factors for some areas were determined using the Radiance daylight simulation model, again at the School of the Built Environment.

These studies resulted in adjustments to the window sizes and positions in some of the spaces and, in particular, a lightshelf was incorporated on the courtyard side of the electrical laboratories to reduce the possibility of glare on the computer screens from the high-level windows and to increase the daylighting levels in the centre of the rooms. The lightshelf was again modelled on the Radiance programme and increased daylight factors were predicted for the centre of the space compared with the initial model of the space without the shelf.

One of the courtyard lightshelves faces north, and on first appearance may seem to be unnecessary. However, the white panels which form the external walls of the electrical laboratory wings in the courtyard are quite reflective (section 5.4) and there is a significant quantity of light reflected from the south-facing courtyard elevation into the north side of the laboratory opposite.

On the south side of the central building the perimeter folds back on itself to form a series of courtyards, thus improving lighting levels in the classrooms and staff offices. The external courtyards may also be used as open-air classrooms in summer months. It could be argued that the extended perimeter required to maximize daylight acts against more traditional ideas of energy conservation by providing a greater external wall area through which heat can be lost in winter. However, the high standard of insulation provided to lower the U-value of the construction in some ways reduces the significance of this issue.

A conflict can also exist between the quantity of glazing required for optimum daylighting and the requirement to reduce solar gain by shading. In addition, if internal blinds or external shades are used to reduce glare and/or heat gain, ventilation by opening windows may be hindered. All these aspects must be considered at an early stage in order to reach the optimum solution.

Figure 13.5 shows light levels for the building.

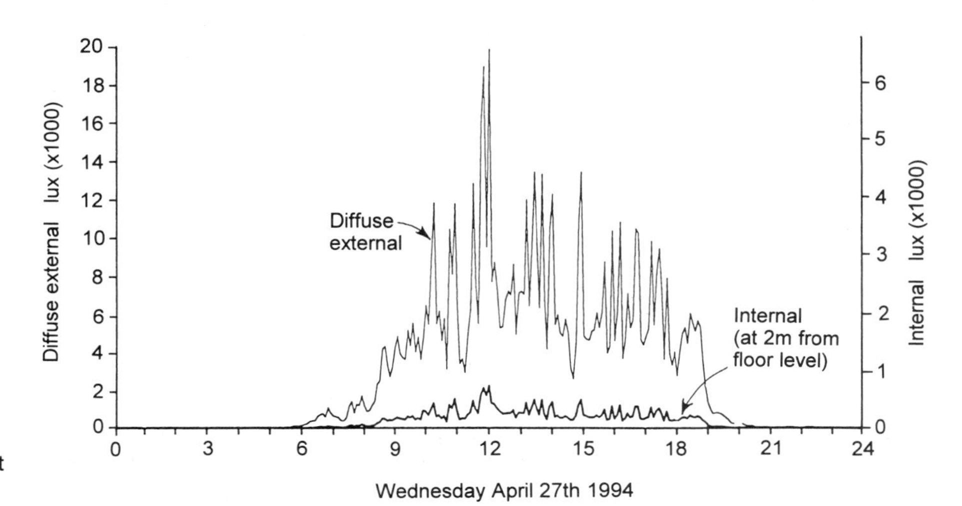

13.5 Measured external diffuse and estimated central concourse internal light levels.

Although there are light level sensors inside and outside the building, those inside will be influenced by any artificial lighting in the vicinity. This is a problem when trying to use internal light level sensors to switch off artificial lighting to save energy when daylight levels are high enough. Thus, only outside sensors are used to switch off artificial lighting in certain spaces. The level at which the artificial lighting is switched off is determined experimentally and iteratively. After the level has been set, the room users then provide feedback; if necessary, the level is altered and the process starts again. This exercise has yet to be undertaken fully within the building. However, initial experiments are being performed. The internal light levels shown in Figure 13.5 were taken while ensuring that the artificial lighting was off.

Artificial lighting

Artificial lighting is generally fluorescent, using either strip lighting or low-energy compact lamps in the classrooms, auditoria, offices, drawing studios, electrical laboratories and circulation spaces. In the offices, teaching rooms and computer laboratories the overhead lighting levels were designed to give 300 lux on the working plane with sufficient sockets available for additional task lighting if required. High-frequency ballasts were incorporated in the fluorescent strip lights to reduce the flicker associated with standard lights of this type as the flicker can affect people with mild epilepsy, particularly when using computer screens. High-frequency ballasts also reduce the energy consumed by the lamp. However, a disadvantage is that high-frequency ballasts are expensive compared to standard ballasts, and this will no doubt remain until they become the industry standard.

In general, the artificial lighting is switched in rows so that those lights nearest the windows can be turned off independently when daylight at the perimeter is adequate. Passive infrared movement detectors can disable the artificial lighting when spaces are unoccupied.

The high spaces of the concourse (Figure 13.6), machine hall and central laboratory have high-bay high-pressure mercury lighting. These lamps were chosen for the machine hall and laboratory to provide the high lighting level of 1000 lux required and also for their efficacy. There is no passive infrared control of the lighting in these two areas as they contain rotating machinery and it could be dangerous to machine operators if the light levels were to alter suddenly.

13.4 Acoustics

The following acoustic issues were addressed during the design phase:

Background noise levels in the auditoria

The air intakes to these spaces are from a busy street. It was agreed with the client that the required environment should provide a balance between the very low noise levels associated with air-conditioned spaces and the noise associated with opening windows in a more traditional classroom environment. The con-

13.6 Concourse space through central building showing daylighting effect.

straints on the solution are cost and space. To remove noise without introducing excessive resistance to air flow requires more space for either acoustically lined plena or acoustic splitters.

Both options were used in the auditoria. The lining is 50 mm Rockwool with a fabric covering to retain the fibres; the splitters are two banks of 500 mm deep, 50 mm wide and 2 m high perforated steel panels filled with Rockwool and suspended in the air intake path with 50 mm gaps in between. In addition, the stacks were lined with 25 mm of fabric-covered Rockwool to reduce potential disturbance from aircraft noise. The L_{A10} noise level (Appendix C) in the street outside the auditoria is 70–75 dBA. Initial sound level measurements have shown a reduction in these levels by 20–25 dBA, giving a background level of approximately 50 dBA just after the attenuators.

Reverberation times in the teaching spaces

The exposed surfaces of brickwork and concrete have low coefficients of absorption and there is therefore a tendency towards high reverberation times. The auditoria reverberation time calculations carried out at the design stage suggested that for a reverberation time of 1.2 to 1.4 seconds at 500 Hz significant amounts of acoustic panelling would be required. This was incorporated on the walls with the soffit left free to expose the concrete, thus taking advantage of its thermal mass. The panels are made of perforated medium density fibre board (MDF) behind which are 50 mm deep Rockwool cavity batts. There is a conflict between the ideal construction for lower peak temperatures and good acoustics, because the exposed surfaces of high mass – such as the brick walls and concrete soffit, which are desirable for their high admittance and can be used as a thermal sink for heat in summer – tend to produce high reverberation times as they have low sound absorption coefficients. A compromise has to be reached that will be acceptable for both criteria.

Noise transfer from inside to outside

The mechanical laboratories shown in Figure 13.7 are sited quite close to a row of domestic houses. Activities such as metal cutting and engine testing can generate internal noise levels in excess of 90 dBA. As a result, the mechanical laboratory ridge vents are open on two sides only (i.e. those facing away from the houses) and the throats of these exhausts are lined internally with a mineral wool quilt in a perforated metal casing. As an additional precaution the high-level gable glazing that faces the houses is triple glazed and fixed.

For this project there was only the standard requirement for acoustic separation between adjacent internal spaces. However, this could be another important

13.7 Mechanical laboratory showing buttress side wall vents with classroom exhaust stack and boiler flue behind.

aspect to be considered in certain situations. At De Montfort each space has its own separate ventilation system with separate exhaust paths, thus there is no risk of noise being carried from place to place by a common stack as often occurs with air-conditioning ductwork serving a number of rooms.

13.5 Controls

One of the main differences between De Montfort's engineering school and a naturally ventilated Victorian building is in the control. From the Victorian era manual controls have been used but often haphazardly. The school, on the other hand, uses a building management system (BMS). Room temperatures and the amount of air entering or leaving the spaces can be controlled using criteria such as:

- fresh air requirement of the occupants (carbon dioxide sensors)
- temperature (individual or averaging temperature sensors in spaces)
- rain (sensor on top of one of the stacks – printed circuit board)
- wind (rotating cup anemometer on top of one of the stacks)
- time (time clock control as part of BMS)
- air movement (flow meters in stacks)
- noise (sound level meter – not yet installed – future provision only)
- fire condition (as sensed by the fire alarm system throughout the building).

It is intended that the building be used as a teaching aid for students and also to further research into naturally ventilated buildings. To this end additional sensors have been incorporated such as relative humidity and temperature sensors within the layers of construction of an external wall, floor and roof. In addition, there are sensors measuring daylight level externally and lighting levels internally. These can be used for practical as well as experimental purposes by programming the computer to switch off the artificial lighting when external daylight reaches a predetermined level.

All control set points are adjustable via the computer but also in each space users can override the BMS using local controls.

The fire alarm overrides all the environmental sensors, shutting air intakes and opening the exhausts. There is also a manual switch for fire-fighting personnel which allows exhausts to be left as dictated by the BMS or closed as required. Sometimes fire officers take the view that exhausts should be opened to allow smoke to escape. At other times they may wish to close the exhausts to starve the fire of oxygen.

In terms of the day-to-day running of the building, there are four basic conditions for which the controls must allow. The auditoria are the most complicated areas and, using them as the example:

1. In summer, if the auditorium is occupied and the inside temperature is higher than the outside temperature (and higher than the internal set temperature) the air inlet and exhaust dampers will be fully open. However, if the internal temperature is lower than the external temperature the dampers will open to a lesser degree based on the reading from the carbon dioxide

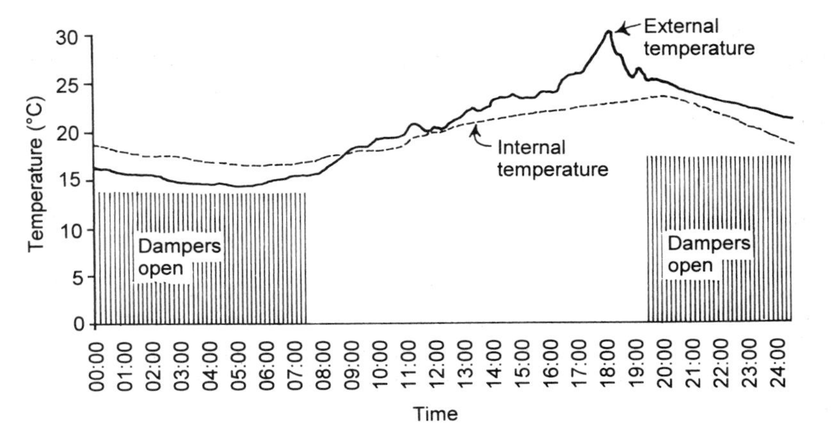

13.8 Recorded temperature data during a heat load test for auditorium 1 on 27–28 June 1994.

sensor. Additionally, if there is rain accompanied by a high wind speed, exhaust dampers will close to a minimal open setting. In extreme winds the air intake and extract dampers will close completely, overriding the carbon dioxide control. The fan will be brought into operation to lower the internal temperature when required.

2. At night, in summer, the external air flows through the space to precool the structure to a preset temperature above the dewpoint (temperature and relative humidity sensors monitored) and no lower than the temperature that would necessitate reheating in the morning for the first lecture. This is set at 17 °C at present. Air flow will be assisted by the fan if necessary.
3. In winter, when the room is occupied, the vents open under the control of the carbon dioxide sensor. As for summer conditions, combinations of wind and rain close the air exhausts and intakes to differing degrees related to severity.
4. At night in winter, or when the rooms are unoccupied, air intakes and exhausts are closed to avoid unnecessary heat loss.

Figure 13.8 shows the outputs from the sensors during a heat load test in one of the auditoria for a 24-hour period where the external conditions reached 30 °C and the internal heat gains were 16 kW. It can be seen that the internal temperature did not rise above 25 °C, which is significantly better than predicted by the early models and calculations (Table 13.4). One likely reason for this is the effect of the wind. The auditoria are currently the subject of extensive monitoring by De Montfort. Results will be published in due course.

13.6 Other aspects

CHP unit

The analysis of the building loads at the design stage showed that large-scale combined heat and power (CHP) plant would not be appropriate as there is little use for large quantities of heat in the summer. A small-scale gas-fired unit was incorporated which produces approximately 40 kW of power and 80 kW of heat. The heat meets the domestic hot water load in the summer months. In winter the CHP unit acts as the lead boiler with a condensing boiler (300 kW duty) as second on line. Two gas-fired high-efficiency boilers (each rated at 300 kW) make up the remaining full heating load. The controls are set to operate to minimize running costs. Thus, if there is no requirement for heat but a requirement for power, the unit will not run and, instead, power is supplied more economically from the grid. Similarly, if there is no power requirement, but a heating requirement the unit will not run and, instead, heat is supplied more economically from the boilers.

The CHP unit does not at present act as a standby generator, although at some time in the future it may be adapted to serve this purpose. The unit has a water-cooled reciprocating engine based on models used in tractors. Heat is reclaimed from the engine jacket, the lubricating oil and the exhaust flue gases to warm the hot water which is fed into the hot water heating system via a water-to-water heat exchanger within the CHP unit. The power generated is run onto the main power intake for the building and is synchronized with the mains electricity supply to ensure that all the power is in phase. If the power of the unit is not being fully used it returns to the mains and feeds the local electricity supply grid. As a safety precaution, therefore, the CHP plant shuts down if the main supply shuts down. This ensures that anyone working on the mains will not find them live through backfeed from the CHP plant.

A further reason for incorporating the CHP unit was to make it available for study by the students of the engineering school.

Solar panels

Some thought was given at an early stage to the incorporation of solar panels to preheat the hot water service. However, the decision to provide a small-scale CHP unit meant that the solar panels would not be a viable option as both items of plant would have been fulfilling a similar function.

Heat recovery

Heat has not been reclaimed from the stack exhausts. Theoretically, it is possible, but the additional resistance imposed by a run-around coil or air-to-air heat exchanger would be likely to impose a requirement for a larger fan in regular operation, and the viability of this would be questionable. In addition, the geometry of the building, and, in particular, the relationship between the air intakes and the exhausts in the auditoria, were not ideally suitable for reclaiming heat from the exhaust and putting it back in at the intake to preheat cold incoming air in winter months. However, heat recovery can be desirable and should be incorporated where appropriate in other designs.

Table 13.5 Approximate capital cost comparison between two comfort cooled auditoria and two naturally ventilated auditoria at De Montfort[4]

	Comfort cooled (£)	*Naturally ventilated (£)*
1. Comfort cooling plant	70 000	–
2. Plant room space	30 000	–
3. External intake louvres	–	9 000
4. Attenuators and controls	–	11 000
5. Plenum duct space	–	5 000
6. Ventilation stacks (4 No.)	–	35 000
7. Attenuators, ventilation terminals and controls to stacks	–	32 000
	100 000	92 000

Costs

One of the main questions that tends to be asked about this building is whether it cost more, or less, to build than a more standard highly massed engineering school with full or part air-conditioning. In part response to this question, Table 13.5 shows some of the figures produced by the quantity surveyors.

The mechanical and electrical services capital costs are obviously much lower, and we also expect running costs to be much lower. Some of the cost of plant has been shifted towards the building.

To reduce costs a specific attempt was made to make aspects of the building relating to natural ventilation serve more than one function. For example, the auditoria stacks provide the main exhausts for air from the auditoria but, in addition, as mentioned above, the structure of the stacks supports the drawing studio roofs above and the glazed tops allow some daylight into the internal parts of the auditoria.

Project principals

Client	De Montfort University, Leicester
Architects	Short Ford & Associates
Services engineers	Max Fordham Associates
Structural engineers	YRM/Anthony Hunt Associates
Quantity surveyors	Dearle & Henderson, London
Landscape architects	Livingston Eyre Associates
Main contractors	Laing Midlands
Mechanical contractors	How Engineering
Electrical contractors	Hall & Stinson

References

1. Data provided by B. Ford (Short Ford & Associates).
2. Lane-Serff, G., Linden, P., Parker, D.J. and Smeed, D.A. (1990) *Laboratory Modelling of the New Engineering School at Leicester Polytechnic*, University of Cambridge Department of Applied Mathematics and Theoretical Physics in conjunction with Cambridge Architectural Research Ltd.
3. Letter of 23 August 1990 from Dr K. Lomas, De Montfort University School of the Built Environment, to A. Short.

Further reading

Bunn, R. (1993) Learning curve. *Building Services*, **15**(10), 20–5.

Hawkes, D. (1994) User control in a passive building. *Architects' Journal*, **10**(199), 27–9.

Stevens, B. (1994) A testing time for natural ventilation. *Building Services*, **16**(11), 51–2.

Swenarton, M. (1993) Low energy gothic: Alan Short and Brian Ford at Leicester. *Architecture Today*, **41**, 20–30.

The Charles Cryer Studio Theatre

CHAPTER 14

This arts centre in Carshalton, London (shown in Figure 14.1), started with a brief for a low maintenance, environmentally friendly design and a specific requirement to provide comfort cooling to the theatre. To achieve an acoustic performance of NR25 (Appendix C) within the theatre space, external noise from the adjacent busy main road had to be excluded and the proximity of local housing meant that performance sound had to be kept in at night. The combination of an acoustically airtight theatre, 250 people and stage lighting heat gains, made ventilation and cooling essential. A range of options was considered including natural and mechanical ventilation, remote-sited air condenser and various water-

14.1 Front elevation.

Table 14.1 Groundwater as a cooling medium

Advantages
- Plentiful quantities of cooling medium
- Minimal plantroom space
- Low maintenance
- Low running cost
- No CFCs or HCFCs
- Simple and traditional equipment and technology
- Recyclable resource

Disadvantages
- Relatively high coil-on water temperature (i.e. the temperature of the water when it enters the cooling coil of an air-handling unit) compared with a conventional air condenser cooling system (i.e. one where the heat is rejected at an air-cooled condenser)
- Information regarding available groundwater and yield may be limited for a particular site
- Generally more expensive than a conventional air condenser cooling system unless 25 years' life cycle costing is considered.

based cooling alternatives. The restricted space in the listed building envelope did not allow suitable plantroom space either inside or outside the building for a traditional air-condenser system. Acoustic enclosures around such a system would have required a total volume of 20–30 m^3.

The solution finally proposed and constructed was to use the rather novel solution of groundwater as a cooling medium. The advantages and disadvantages of groundwater are shown in Table 14.1.

There are many potential sources of groundwater, and many methods of extraction and ways in which the water can be used. The first step was to survey the surrounding area, study geological maps and discuss the options with the National Rivers Authority (NRA),[1] local water authority and the British Geological Survey (BGS).[2]

The NRA and BGS, with their extensive knowledge of a given area, can quickly determine the probability of suitable groundwater sources.

This research will generally provide an overview of the local geology, areas of surface water, underground water, location and extraction capacity of the local boreholes and the seasonal and secular (i.e. yearly) reliability (variation) in water levels.

For the Arts Centre the most suitable and reliable method of extraction was considered to be a deep borehole to tap the water from an underground aquifer. Substantial amounts of surface water appeared to be available locally but the source was unreliable.

The NRA and BGS data suggested that an abstraction borehole drilled 50 m deep through the chalk strata and into the underground aquifer would provide a reliable water table well (Figure 14.2).

Similar boreholes in the area suggested that the natural water level in such a borehole would be within 5 m of the ground level with seasonal variations of up to 1 m. Yield was expected to be more than adequate and a water temperature of

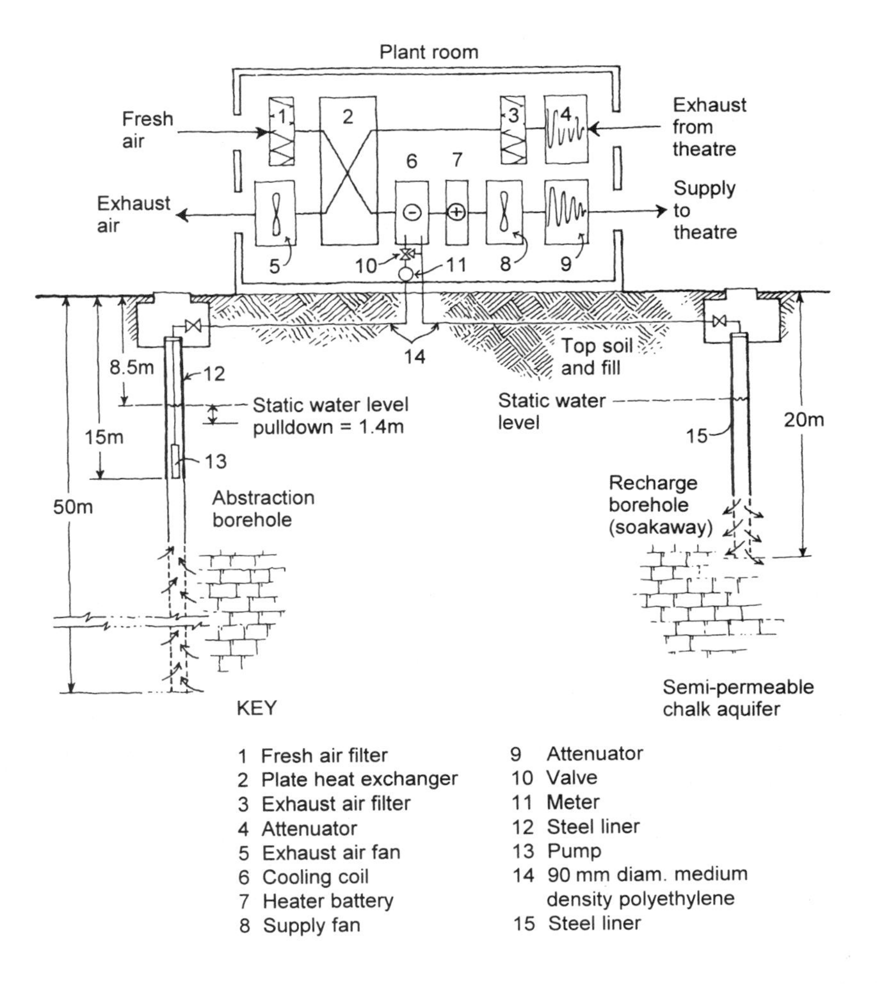

14.2 Borehole cooling and ventilation schematic.

10 °C was normal for similar 50 m deep boreholes in the area – at this depth the temperature appears to be constant. The strata of the chalk aquifer generally provides clean, clear potable water. Although the water is very hard (i.e. has a high content of calcium and magnesium bicarbonates, in this case 220 mg/l, when expressed as $CaCO_3$), the water in the heat exchanger would be less than 20 °C, which should not cause fouling of the waterways.

To conserve and recycle this natural resource it is necessary to ensure that the water is not contaminated in use and that it is returned to the aquifer. This is done, in part, by separating the supply and return water so that the return water does not raise the temperature of the supply water. Normally, a separation of 100 m is recommended, but this was not feasible in this case and so a second 20 m deep borehole (soakaway), which returns water to the aquifer, was located 40 m distant from the first. To provide additional security the boreholes are set at different levels in the aquifer and with the abstraction borehole on the upstream side of the nominal direction of flow of the aquifer.

Table 14.2 Theatre ventilation system performance data

Full speed fan volume	4.2 m^3/s
Fan power	7.5 kW
Air on-coil	26 °C dry bulb; 19 °C wet bulb
Air off-coil	13.8 °C dry bulb; 13.4 °C wet bulb
Cooling water on-coil	11 °C
Cooling water off-coil	18 °C
Cooling water flow rate	2.8 l/s
Cooling coil rows	8
Cooling duty	83 kW
Maximum capacity (persons)	250
Ventilation rate per person (maximum)	17 l/s
Air changes per hour (maximum)	12

A multi-stage borehole pump installed inside the borehole (and so out of sight and requiring no plantroom space) raises water from the borehole to the air-handling unit, which is 10 m above ground level.

Although the current water resource is considered to be plentiful and recycled, the water extracted is metered and the annual quantities 'borrowed' are restricted by the NRA. In order to minimize the pumped water demand and therefore energy used and running costs, suitable measures have been taken to reduce the cooling demand.

The building relies on the heavy thermal mass of the fabric, precooling and free cooling (Chapter 13) to reduce the cooling loads. Free cooling uses the external supply air to help to cool the building (without running the borehole water pump) when the external air temperature is lower than the space design temperature. Two speed ventilation fans ensure that fan power can be minimized when the full cooling capacity is not required. When the full cooling load is expected, the management can use the manual override controls to precool the theatre to 18 °C and rely on the thermal mass of the exposed heavy building fabric to maintain comfort conditions for a limited period.

The mechanical ventilation system is a full fresh air (no recirculated air or bypass dampers), two-speed, ducted supply air and exhaust system with a heater battery, cooling battery, silencers, filters and plate heat exchanger for heat recovery. Fresh air enters the system from the north side of the building where the tree-lined courtyard air is cooler and less polluted and the ambient noise is lower. Fire dampers, bypass and air flow regulation dampers were designed out of the ventilation system to reduce maintenance. The measures taken to achieve low maintenance and reliability do, however, compromise the ideal conditions required for low cooling loads. The permanently in-line plate heat exchanger reduces the potential for free cooling, but in practice this is not a problem. Decorative finishes inside the building reduce the thermal mass available by insulating the heavy structure and building fabric from the space. Nevertheless, the measures taken reduced the design cooling load to 395 W/m^2. The system performance data is given in Table 14.2.

With the peak cooling water flow rate established from the cooling duty as 2.8

Table 14.3 Borehole statistics

	Abstraction borehole	*Recharge borehole*
Nominal diameter (mm)	300	375
Depth (m)	50	20
Pumped flow (m^3/h)	10	10
Pulldown[a] (m)	1.4	N/A
Steel liner depth (m)	15	15
Static water level below ground (m)	8.5	8.4
Pump inlet level below ground (m)	15	N/A
Variation in static water level (m)	1	1

[a]Pulldown is the difference between the water level in the borehole when the pump is running and when it is not.

l/s, the borehole diameters could be sized to achieve the necessary abstraction and recharge duties. The NRA and BGS can provide the necessary background information for this assessment. Table 14.3 lists the relevant borehole statistics.

The abstraction borehole has a mild steel lining tube (10 mm thick) for the top 15 m. The inlet to the pump is 15 m below ground level with a steel discharge pipe running to a flanged base plate in the abstraction manhole.

All pipework from the plantroom to each borehole is laid to fall to the boreholes and provided with a vent at the highest point in the plantroom. The discharge water from the cooling battery returns to the recharge borehole by gravity. When the borehole pump is not running, the water in the supply pipework drains down into the borehole so preventing any danger from frozen pipes, leaking joints or stagnant water.

Fully automatic or manual control is available to the theatre management. Under automatic or manual cooling mode, averaging thermostats will operate the borehole pump when the room air temperature exceeds a predetermined level. A motorized three-port valve in the borehole supply pipework then controls the volume of borehole water available to the cooling battery. A further thermostat in the supply air ductwork controls the motorized three-port valve to ensure that the minimum supply air temperature does not fall below 14 °C.

Each borehole cost approximately £6000, but the complete package – including boreholes, pipework, pump, controls and manholes – is nearer £30 000. With 83 kW of cooling this is over £360 per kilowatt. However, the installed system could easily provide 160 kW of cooling for little more than the cost of a higher duty pump, thus bringing costs down to about £200/kW.

If a traditional air condenser package had been feasible, the installed capital cost would have been approximately £15 000. Such a comparison is not strictly accurate, of course, as there are many other costs and considerations that should be taken into account, such as maintenance costs, running costs and the effect that the relatively high cooling medium water temperature has on the sizing of the air-handling unit.

Underground aquifers can provide water at temperatures as low as 10 °C but by the time this reaches the cooling coil of an air-handling unit the water on-coil temperature is likely to be about 12 °C. It is then difficult to achieve air temperatures off the coil lower than about 14 °C. As a result, the air-flow rates, air-handling unit size, fan size and ductwork may need to be increased to achieve the required cooling duty.

Maintenance costs on the borehole have proved to be extremely low. The pump and the motorized valve are the only moving parts. Borehole water can deposit a slime within the pipework and cooling coil. To allow any such residue to be cleaned out the cooling coil is provided with a flushing facility although it has not yet been necessary to use it.

Running costs are low as only the 2.5 kW pump, 65 mm motorized valve and the controls require power. The coefficient of performance (COP) of the system is very good. With the fans running at full speed the electrical power absorbed by the fans, pump and controls is approximately 12 kW, and this provides a cooling output of 83 kW.

A number of variations and modifications to this scheme were considered (e.g. run-around coils, recirculation dampers, evaporative cooling, water-to-water chiller) to improve performance, efficiency and/or energy consumption but all these variations would have had an adverse effect on either capital cost, maintenance, simplicity or reliability.

Summary

Groundwater is an environmentally friendly answer to many cooling needs and this system at the Charles Cryer Studio Theatre has proved to be very simple, reliable and economical to operate and maintain. If groundwater is available locally, and if demand is not oversubscribed, such cooling may well be a feasible option.

Project principals

Client	London Borough of Sutton
Architects	Edward Cullinan Architects
Services engineers	Max Fordham & Partners
Structural engineers	Jampel Davison & Bell
Quantity surveyors	Dearle & Henderson
Main contractors	Eve Construction
Mechanical contractors	J.W. Stubberfield & Sons Ltd
Electrical contractors	RTT Engineering Services

References

1. National Rivers Authority, Thames Region, Kings Meadow House, Kings Meadow Road, Reading, Berkshire.

2. British Geological Survey, Maclean Building, Crowmarsh Gifford, Wallingford, Oxfordshire OX10 8BB.

Further reading

Todd, D.K. (1959) *Ground Water Hydrology*, Wiley, Tokyo.
Wright, P. (1992) Dramatic effect. *Architects' Journal*, **16**(195), 36–45.

CHAPTER 15 Sutton House

15.1 The building

Sutton House in Hackney, shown as built in Figure 15.1, is the oldest building in east London. It originated in the Tudor era and has witnessed many periods, each of which has left its mark. This is most apparent in the façade, interior panelled rooms and additions. The building has recently been restored and rejoins the community as an educational and recreational asset and home to the National Trust youth theatre.

15.1 Sutton House as built in 1535.

15.2 The linenfold panelled room after restoration.

The building restoration was an interesting challenge because of the demands that multi-purpose usage, which was required to make the project viable, placed upon the environment. The building was to house offices, a café/bar, a gallery and conference facilities and to be an educational museum. The six panelled rooms, varying in period from Tudor linenfold (Figure 15.2) to Georgian, required careful environmental management to preserve the panelling. This need had to be reconciled with the demands for higher temperatures required for the comfort of visitors and staff. Because of the historic importance of the structure it was not possible to alter it significantly. One result of this was limited available space for services.

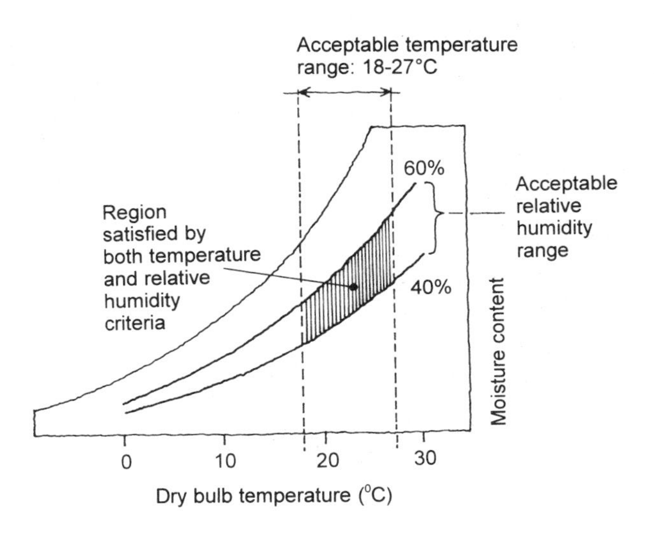

15.3 Environmental requirements for occupancy and panelling.

Unlike most commercial or restoration projects, no one factor had carte blanche in defining the brief for the environmental services. Occupancy demands vary, giving a range of temperature between, say, 18 and 27 °C, whereas the wood panelling is best preserved by a constant relative humidity. In this instance, however, it was agreed that a range of relative humidity of 40–60% would be acceptable for the panelling. The envelope that both conditions define simultaneously is small (Figure 15.3). Traditionally air-conditioning is used to provide the necessary conditions, but the budget did not cover it. We therefore examined what could be achieved without air-conditioning.

The 40–60% relative humidity range (rather than an ideal 55%) had been agreed with the client as suitable for the panelling as long as changes occurred only slowly, i.e. over weeks or months. It was thought that the occurrences of intermittent occupancy (when higher required heating temperatures drive down the relative humidity) and low ambient relative humidity were infrequent and brief enough so as not to be harmful to the panelling. This is because the building fabric, by absorbing and emitting moisture, dampens any short-term effects and because the condition of the panelling depends more on long-term conditions than transient states. It should be noted that this does impose limitations on the usage of the rooms, as consistent over-use of the rooms, in periods of low ambient relative humidity, would alter the long-term conditions.

Given allowable ranges for occupancy comfort and panelling preservation, we were able to analyse typical weather data over a period of a year and identify a 'heating-only' strategy that had the potential to provide acceptable conditions (Figure 15.4).

A building with a heating-only strategy ultimately responds to and follows outside conditions. In this instance we were interested in the varying external moisture content. The building itself evens out this varying condition and also introduces a delay between the effect of external conditions on internal ones. A controlled heating strategy can aid in this process by increasing or decreasing the

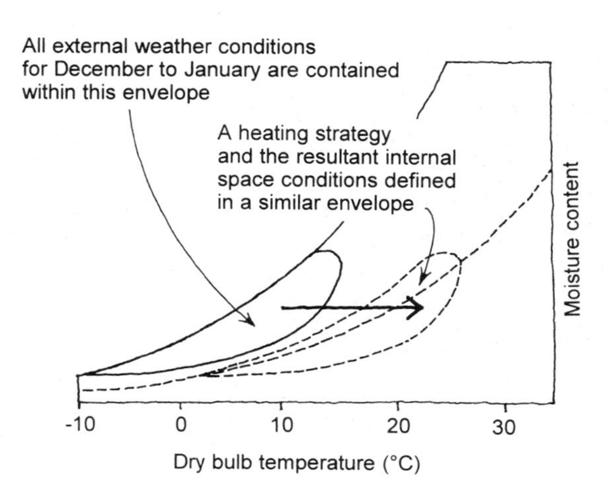

15.4 Winter weather represented as an envelope of conditions and the effect of a heating strategy.

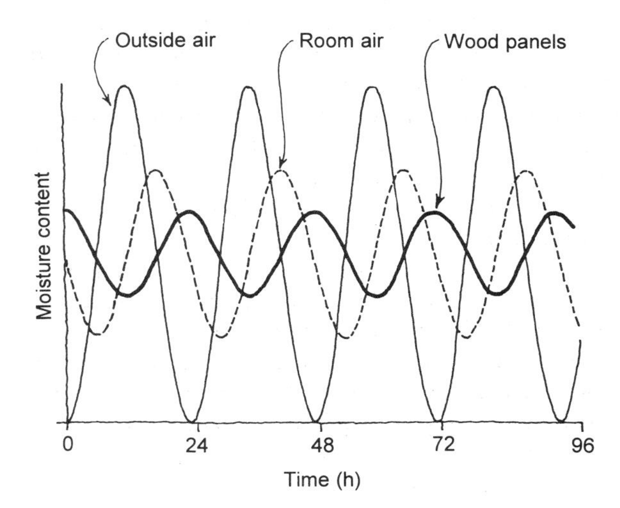

15.5 A comparison of external, room and wood panel moisture contents.

room temperature and thus increasing the damping effect. This is indicated in Figure 15.5.

Of course, a heating-only strategy cannot add moisture and there will be occurrences of low relative humidity, which, if prolonged, will require separate action such as introducing portable humidifiers or not using the rooms. However, these situations are not expected to happen often. It is also worth noting that when many people occupy a room they produce moisture, and this addition of moisture means that the temperature of the space can be raised without having to lower the relative humidity level (compared to that of the space when it is unoccupied).

15.6 The restored linenfold window reveals (natural convectors are integrated into the window seats).

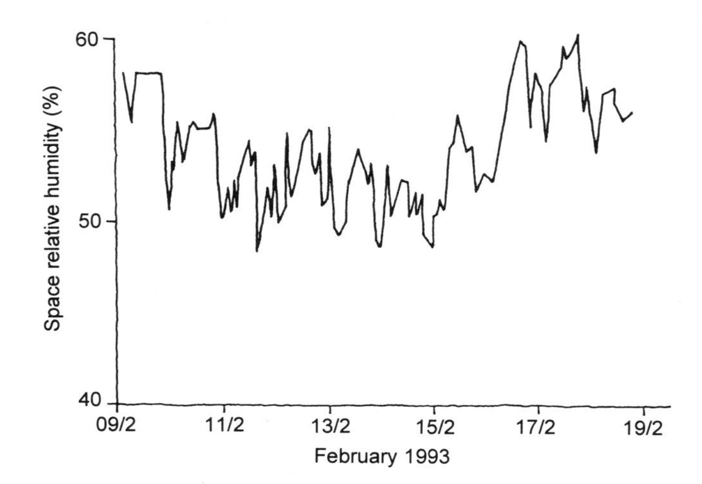

15.7 Space relative humidity in the linenfold panelled room.

We have also found that although people have added moisture, on particularly cold days this has not had enough of an effect to allow the temperature to reach comfort level and, consequently, some people have found the rooms chilly.

A compensated low-temperature hot water heating system serving high-capacity natural convectors that are carefully concealed in the window seats (Figure 15.6), with individual room on–off valves and controlled by a building management system (BMS), was chosen as being most suitable. The convectors offered a quick warm-up time to allow rooms to be used as required. This system provided a reliable solution that did not require large service voids or the use of fans, ducts or air-conditioning. It also offered inexpensive maintenance.

The BMS controls and monitors each room individually. While a room is unoccupied the BMS controls the space relative humidity indirectly on the basis of a space air temperature criterion of 6 °C above outside air temperature. This has proved to produce good control of relative humidity in other historic buildings.[1] When the room is to be occupied a local manual override switches the control criterion to a higher temperature which is determined from the current space relative humidity and the minimum humidity requirement. This means that the BMS is controlling for the majority of the time from temperature sensors. With the possibility of infrequent maintenance, this method was considered more robust than solely relying on relative humidity sensors which require checking approximately every six months. At night the controls allow the space temperatures to drop as low as 5 °C, thus minimizing the risk of very low relative humidities.

The project has been monitored since its completion in 1992. During the winter of 1992–93 (which admittedly was not severe), good control of relative humidity was maintained with very few occurrences of values below 40%. Figure 15.7 shows the results for the linenfold room for a short period: as these figures are based on space air measurements we expect that the actual conditions for the panelling itself are smoother.

An interesting comment from the client during this period was that he associated warmer conditions in the panelled rooms with wet weather (which showed that the strategy was working). This is because the system allows higher occupied temperatures as more moisture is introduced in the space.

During the monitoring period observations led to a number of conclusions. The first was that one room experienced a quicker decay of relative humidity during the night period. This was attributed to greater cross ventilation, as, at the time of monitoring, this room in particular had 'cracks' on both the front and rear elevations.

We also found that it was critical to calibrate humidity sensors carefully, allowing them to adjust to the site at installation and during the commissioning period.

The building was gently allowed to dry out over the first six months, with limited occupancy in the panelled rooms. The control criteria were also gently imposed. The shrinkage in the panelling was very visible by observing the 'tide mark' of the surface treatment on the individual panel edges retracting during the drying-out period. It was essential at this time to ensure that the panels were free to move.

On the larger softwood Georgian panelling the edge movement was as much as 5 mm on each side. Only one Georgian panel suffered damage; this was caused by a restrained edge stopping even shrinkage and causing the panel to split violently.

This project demonstrates how occupancy and conservation can be designed to coexist in appropriate applications, while utilizing only simple equipment. The use of expensive air-conditioning is avoided along with its attendant problems of maintenance, space, cost and detrimental effects on the environment.

There is no doubt that air-conditioning, as a solution, would have provided tight relative humidity control and comfortable temperatures all the time, yet air-conditioning is all too often chosen as an easy solution to a difficult problem, rather than as a last resort.

The solution used in Sutton House is a compromise that exploits the nature of the building. It has limitations in meeting expectations of comfort and times of use, but for the most part it has proved to be very satisfactory.

Project principals

Client	The National Trust
Architects	Richard Griffiths Architects, Julian Harrap Architects
Services engineers	Max Fordham & Partners
Structural engineers	Hockley & Dawson
Quantity surveyors	Sawyer & Fisher
Main and electrical contractors	Loe & Co.
Mechanical contractors	Willsmore & Son

Reference

1. From data gathered and interpreted by Bob Hayes of Colebrook Consulting for the National Trust.

Appendices

Environmental data and the psychrometric chart

APPENDIX A

A.1 Solar data

Hours of sunshine

Figure A.1 and Table A.1 give data on sunshine in London.

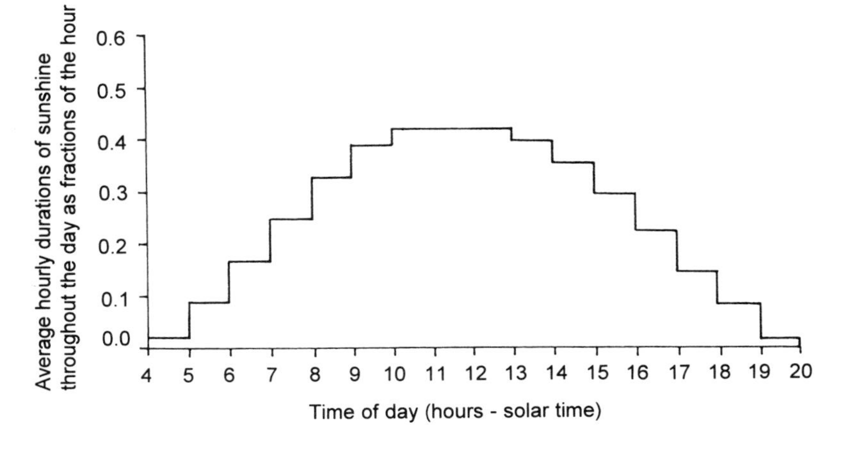

A.1 Hourly durations of sunshine for the whole year for London.[1]

Table A.1 Sunshine hours[2]

Period[a]	*Average hours of sunshine per day in London*
1. Spring 4 a.m.–8 p.m.	4.78
2. Summer 4 a.m.–8 p.m.	6.28
3. Autumn 6 a.m.–6 p.m.	3.51
4. Winter 7 a.m.–5 p.m.	1.71
5. Year	4.07

[a] Spring = March–May; Summer = June–August; Autumn = September–November; Winter = December–February.

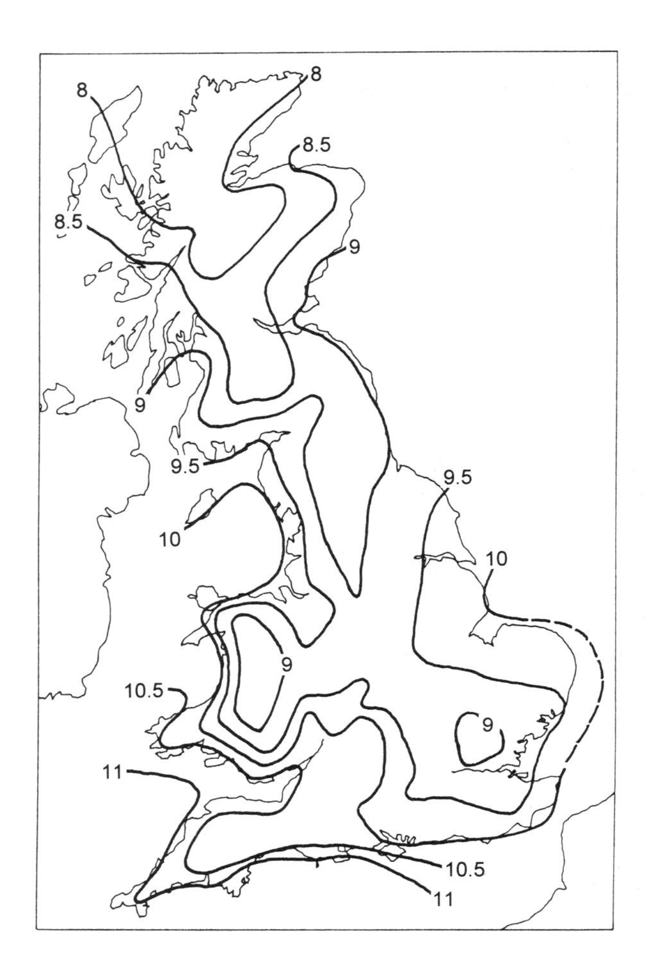

A.2 Annual mean of daily total global solar irradiation on a horizontal surface in MJ/m^2.[3]

Incident solar radiation in the United Kingdom

Figure A.2 gives solar radiation data.

Solar altitudes and azimuths

Extensive data is available on solar position – Table A.2 gives a very basic introduction.

Table A.2 Approximate solar altitudes and azimuths at 52 °N[4]

Date and time	*Altitude (deg.)*	*Azimuth (deg.)*
December 21		
0900	5	139
1200	15	180
1500	5	221
March 21 and September 22		
0800	18	114
1200	38	180
1600	18	246
June 21		
0800	37	98
1200	62	180
1600	37	262

A.2 Daylighting

Table A.3 gives data on daylighting levels.

Table A.3 Illuminance at Kew, London[5]

Illuminance interval (lux)	*Percentage of year[a] diffuse illuminance is above lower limit of range (%)*	*Percentage of year[a] total illuminance[b] is above lower limit of range (%)*
0–1 000	100.0	100.0
1 000–5 000	94.3	94.4
5 000–25 000	84.1	84.6
25 000–50 000	30.7	48.6
50 000–75 000	1.3	23.8
75 000–100 000	0.0	8.0

[a]Based on a standard working year of 0900–1730, British Standard Time, April to October inclusive.
[b]Total illuminance is that due to light from the Sun and sky; diffuse illuminance is that due from the sky alone.

A.3 Temperature data

As for solar radiation, extensive temperature data is available. Table A.4 gives a very limited selection.

In a hot summer in England, such as that of 1994, peak temperatures can be 25–30 °C. The hottest July day since accurate recording began saw a peak temperature of 35.9 °C at Cheltenham in 1976 – for August, the similar peak was 37.1 °C in 1990, also at Cheltenham.[7]

In 1994 a fairly typical pattern for a hot summer's day was a minimum of 18 or 19 °C at night, rising to, say, 28 °C or so at 3 or 4 p.m. Judging again from the summer of 1994, winds were light to moderate (let us say, very approximately, 0.3–5.0 m/s) during both periods. We shall return to this below.

Table A.4 Daily mean dry bulb temperatures at Wisley, London, 1941–70[6]

Month	*Temperature (°C)*	*Month*	*Temperature (°C)*
January	3.5	July	16.9
February	4.0	August	16.5
March	6.3	September	14.4
April	9.1	October	10.9
May	12.1	November	6.8
June	15.2	December	4.5

Average for period 1941–70 = 10 °C

A.4 Wind

Table A.5 gives the Beaufort scale, which is useful in developing a personal sense of wind speeds.

Wind data is available from a number of sources.[8,9] The UK is one of the windiest countries in the world, and even in the southern parts of England the mean wind speeds are above 4–5 m/s 50% of the time, as can be seen in Figure A.3.

Figure A.4 shows typical wind roses for Kew, London. The lengths of the lines are proportional to the amount of time the wind comes from the direction indicated.

An area of particular interest for designers of assisted naturally ventilated buildings is the frequency of low wind speeds and high temperatures, because these conditions are most likely to lead to discomfort. Little published analysis of available data appears to have been carried out. We have examined 10 years' of data from Kew, London[12] and found that June and July have fairly high mean temperatures (say 16–17 °C), high maximum temperatures (say 28–31 °C) and high monthly total solar radiation figures (say about 140 kWh/m^2 of horizontal surface). Mean wind speeds for June and July were 3.3 and 3.6 m/s,

Table A.5 Beaufort scale

Beaufort number	*Windspeed (m/s[a])*	*Description*	*Land condition*	*Comfort*
0	0–0.5	Calm	Smoke rises vertically	No noticeable wind
1	0.5–1.5	Light air	Smoke drifts	
2	1.6–3.3	Light breeze	Leaves rustle	Wind felt on face
3	3.4–5.4	Gentle breeze	Wind extends flags	Hair disturbed, clothing flaps
4	5.5–7.9	Moderate breeze	Small branches in motion, raises dust and loose paper	Hair disarranged
5	8.0–10.7	Fresh breeze	Small trees in leaf begin to sway	Force of wind felt on body
6	10.8–13.8	Strong breeze	Whistling in telegraph wires, large branches in motion	Umbrellas used with difficulty Difficult to walk steadily Noise in ears
7	13.9–17.1	Near gale	Whole trees in motion	Inconvenience in walking
8	17.2–20.7	Gale	Twigs broken from trees	Progress impeded Balance difficult in gusts
9	20.8–24.4	Strong gale	Slight structural damage (chimney pots and slates)	People blown over in gusts
10	24.4–28.5	Storm	Seldom experienced inland. Trees up-rooted, considerable structural damage	

[a]Measured 10 m above sea or ground level.

respectively, compared to a yearly mean of about 3.9 m/s. Examination of one particularly hot day in July with a minimum temperature of 15 °C at 4.00 a.m. and a peak of 30.6 °C at 4.00 p.m. showed a mean wind speed of 1.6 m/s with a minimum of 0.0 m/s and a peak of 4.9 m/s. The day in fact had a period of six continuous hours from midnight to 6.00 a.m. when the mean wind speed was less than 0.1 m/s; on the other hand, during the hottest period from noon to 6.00 p.m., the mean wind speed was 4.2 m/s. Obviously, much more analysis is needed. Cautious designers should probably plan on low wind speeds and high temperatures for short periods of interest, say, four hours. As the period of interest lengthens, it would appear reasonable to use somewhat higher wind speeds and lower temperatures.

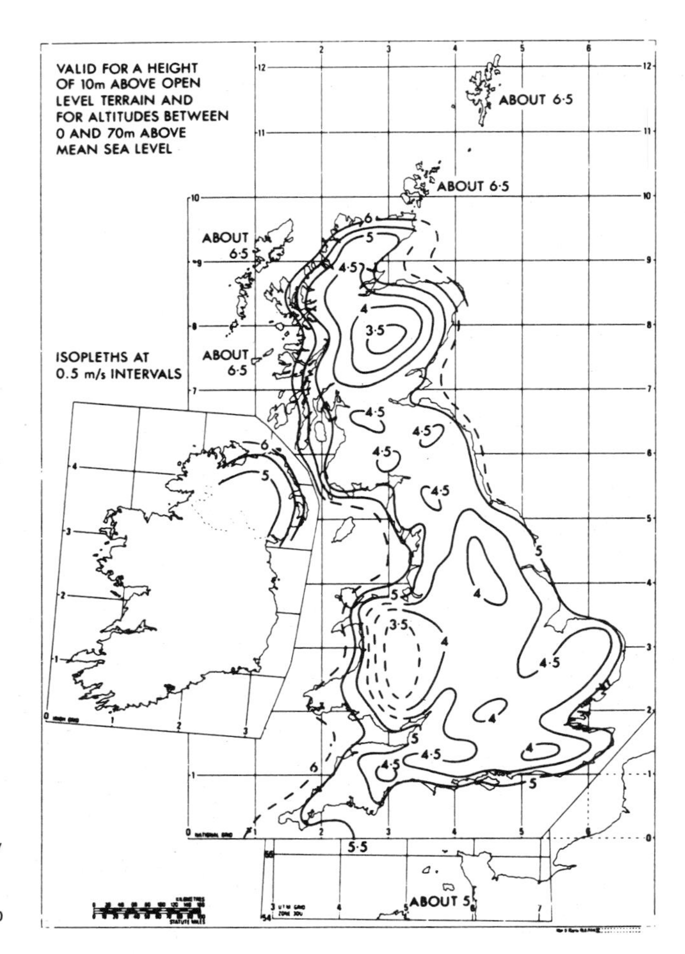

A.3 Isopleths of hourly mean wind speeds exceeded for 50% of the time over the UK.[10]

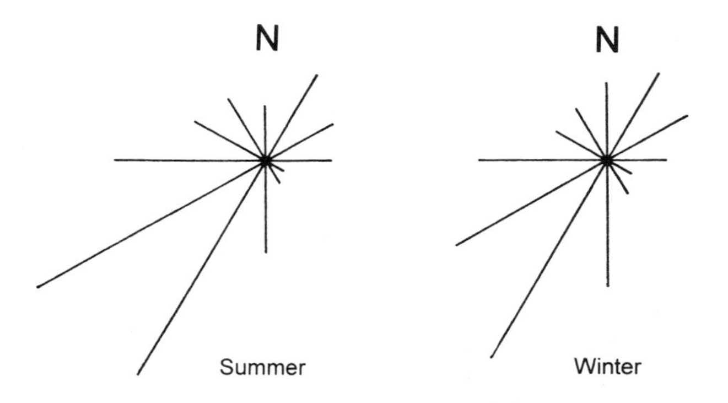

A.4 Wind roses for Kew, London.[11]

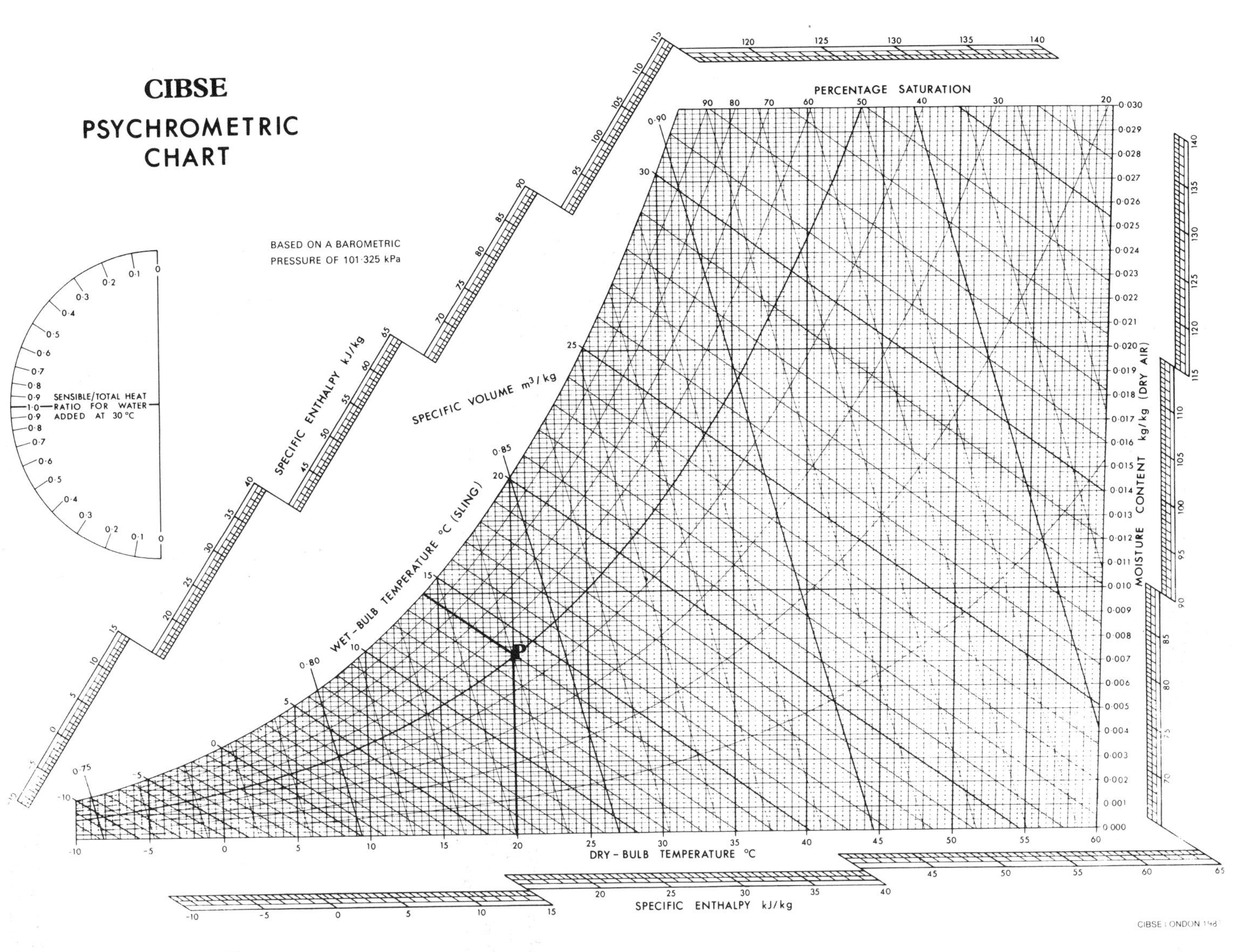

A.5 Psychrometric chart.[13]

A.5 Psychrometric chart

Figure A.5 shows the psychrometric chart, which relates temperature and moisture content.

The dry-bulb temperature is that measured by an ordinary thermometer. The wet-bulb temperature is measured by a thermometer with a wetted sleeve and so its readings are affected by the moisture content of the air (which is shown on the right-hand axis). The percentage saturation is the amount of moisture in the air at a given temperature compared to the amount of moisture in saturated air at the same temperature. Within the range of conditions normally encountered in buildings it can be taken as virtually the same value as the relative humidity. Relative humidity is defined as the ratio of the partial pressure of moist air at a given temperature to the partial pressure of the water vapour in saturated air at the same temperature.

As an example of the use of the chart, point P shows a condition of 20 °C dry bulb and 14 °C wet bulb, which is 50% saturation or, for our purposes, relative humidity. The moisture content of the air in this case is about 7 g/kg of dry air. The dewpoint, i.e. the point of 100% saturation, is just under 10 °C.

Table A.6 gives the frequency (on an annual basis) of percentage saturation ranges.

Table A.6 Approximate annual distribution of percentage saturation for London (based on 20 years' data)[14]

Percentage saturation (%)	*Percentage of year (%)*
90–100	30
80–89	24
70–79	17
60–69	14
50–59	9
40–49	4
30–39	2

References

1. Ne'eman, E. and Light, W. (1975) Availability of sunshine. BRE CP 75/75. BRE, Garston.
2. Ibid., pp. 8–11.
3. Anon. (1986) *CIBSE Guide A2: Weather and Solar Data*, CIBSE, London.
4. Ibid., pp. A2–56.
5. Hunt, D.R.G. (1979) *Availability of Daylight*, BRE, Garston.
6. Values of Climatological Averages at Wisley (undated), Meteorological Office, London.
7. Nuttal, N. (1994) Sun soaked Britain will soon be put in the shade. *The Times*, 12 July.

8. See reference 3.
9. Anon. (1993) *ASHRAE Handbook – Fundamentals: Weather Data*, ASHRAE, Atlanta.
10. See reference 3.
11. Anon. (1971) *IHVE Guide A2*, IHVE, London.
12. Data from J. Littler of University of Westminster; analysis by R. Thomas, Max Fordham & Partners
13. Anon. (Undated) *Psychrometric Chart for Dry-bulb Temperatures: 10°C to 60°C*, CIBSE, London.
14. Holmes, M. and Adams, S. (1977) Coincidence of dry and wet bulb temperatures. TN 2/77. BSRIA, Bracknell. (Data derived from this source by the author.)

APPENDIX B

Calculation procedures

B.1 U-value calculations

The U-value is the rate of heat flow per unit area from the fluid (usually air) on the warm side of an element to the fluid (again usually air) on the cold side. The procedure for U-value calculations is given in numerous references[1,2,3] along with thermal conductivities for various materials. Values of thermal conductivity vary somewhat according to the reference consulted. Manufacturers will provide the most accurate test data for their products.

Here we shall simply give one example of a calculation for a wall at the De Montfort University Queens Building (Figure B1; Chapter 13).

1. Construction
   - 190 mm dense concrete block (*NB* dense block was used because it is a load-bearing element in this four-storey all masonry construction building)
   - 100 mm Rockwool cavity batts
   - 100 mm brick (*NB* 100 mm is nominal and suitable for our purposes. The actual brick width is 102.5 mm).
2. Internal surface resistance: use 0.12 m^2 K/W

B.1 The wall under construction.

External surface resistance: use 0.06 m^2 K/W.

3. Thermal conductivity:
   dense concrete block: 1.63 W/m K
   Rockwool: 0.035 W/m K
   brick: 0.84 W/m K.
4. To calculate the thermal resistance of a building element:

$$\text{Resistance}\ \frac{(\text{m}^2\ \text{K})}{\text{W}} = \frac{\text{Thickness (m)}}{\text{Conductivity (W/m K)}}$$

5. Add the resistances of all elements:
   Internal surface resistance = 0.12 m^2 K/W

$$\text{Resistance of block} = \frac{0.19\ \text{m}}{1.63\ \text{W/m K}} = 0.12\ \text{m}^2\ \text{K/W}$$

$$\text{Resistance of Rockwool} = \frac{0.1\ \text{m}}{0.035\ \text{W/m K}} = 2.86\ \text{m}^2\ \text{K/W}$$

$$\text{Resistance of brick} = \frac{0.1025\ \text{m}}{0.84\ \text{W/m K}} = 0.12\ \text{m}^2\ \text{K/W}$$

   External surface resistance = 0.06 m^2 K/W

   Sum of resistances = 3.28 m^2 K/W

6. The U-value is the reciprocal of this:

$$\frac{1}{\text{Sum of resistances}} = \frac{1}{3.28\ \text{m}^2\ \text{K/W}} = 0.30\ \text{W/m}^2\text{K}$$

B.2 Daylighting calculations

A number of references deal with estimating daylight in buildings and the calculation of the average daylight factor (ADF).[4,5] The ADF (Chapter 5) for a side-lit interior is given by

$$\text{ADF} = \frac{TA_w\theta}{A(1 - R^2)}$$

where T is the diffuse light transmittance of the glazing including the effects of dirt, blinds, obstructions and coverings; A_w is the window area (m^2); θ is the vertical angle subtended at the centre of the window by unobstructed sky; A is the total area of indoor surfaces (ceiling, walls and floor, including glazing); and R is the area-weighted average reflectance of ceilings, walls and windows. We can apply this to the very simple case shown in Figure B.2.

Let us assume that T = 0.75 and

Reflectance of the ceiling = 0.7
Reflectance of the wall = 0.5
Reflectance of the window = 0.1
Reflectance of the floor = 0.3

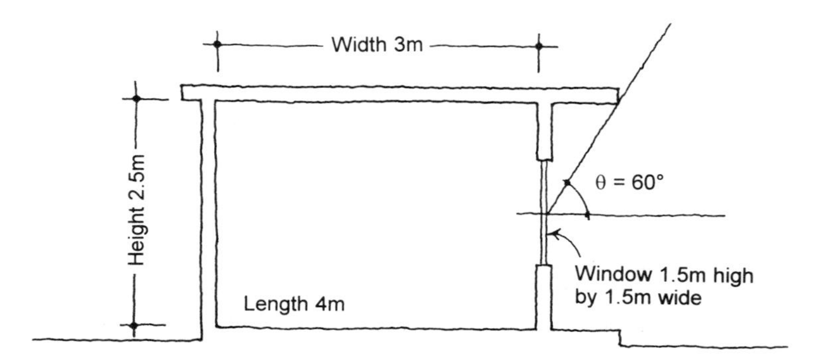

B.2 Daylighting for side-lit interiors.

The total area of the room is 59 m^2. The average reflectance is area weighted in the following way:

$$\begin{aligned}
(R \text{ side wall}) \times (\text{Area side wall}) &= (0.5)(3)(2.5) &= 3.75\\
(R \text{ side wall}) \times (\text{Area side wall}) &= (0.5)(3)(2.5) &= 3.75\\
(R \text{ back wall}) \times (\text{Area back wall}) &= (0.5)(4)(2.5) &= 5.00\\
(R \text{ front wall}) \times (\text{Area front wall}) &= (0.5)(10-2.25) &= 3.88\\
(R \text{ window}) \times (\text{Area window}) &= (0.1)(2.25) &= 0.23\\
(R \text{ ceiling}) \times (\text{Area ceiling}) &= (0.7)(4)(3) &= 8.4\\
(R \text{ floor}) \times (\text{Area floor}) &= (0.3)(4)(3) &= 3.6\\
&\text{Total} &= 28.61
\end{aligned}$$

then,

$$\mathbf{R} = \frac{28.61}{59} = 0.48$$

and

$$\text{ADF} = \frac{0.75(2.25)(60)}{59[1-(0.48)(0.48)]} = 2.2\%$$

Thus, in conditions of a standard overcast sky of 5000 lux, the average light level would be 5000 × 2.2% = 110 lux.

References

1. Anon. (1980) *CIBSE Guide A3: Thermal Properties of Building Structures*, CIBSE, London.
2. Burberry, R. (1992) *Environment and Services*, Longman, Harlow.
3. Littler, J.G.F. and Thomas, R.B. (1984) *Design with Energy*, Cambridge University Press, Cambridge.
4. Anon. (1994) CIBSE code for interior lighting. CIBSE, London.
5. Anon. (1986) Estimating daylight in buildings: Part 2. BRE Digest 310. BRE, Garston.

Acoustics

APPENDIX C

In Chapter 2 we saw that sound levels are measured according to a logarithmic scale of decibels and that weighting to allow for the response of the human ear gives dBA values. A decibel is defined as 10 times the logarithm to the base 10 of a ratio of two powers, thus:

$$\text{dB} = 10 \log_{10} \frac{\text{Power}}{\text{Reference power}}$$

For sound power levels the reference power is taken as 10^{-12} watts. Therefore, if a source is 100 times more powerful than this it will measure as 20 dB, and if it is 1000 times more powerful it will measure as 30 dB.

Sound power is the power emitted by an acoustic source, say a loudspeaker in a room, and is measured in watts. The resulting loudness, or more precisely sound pressure level (Table C.1), in the room is dependent not only on the loudspeaker but on the characteristics of the room (and its furnishings).

Most noises are, of course, variable. To deal with this, sound pressure levels can be examined over time – if a high level is infrequent it is likely to be more acceptable. An L_{10} level is that value of noise exceeded 10% of the time; L_{A10} is that value when measuring in dBA. It is also possible to average the sound. $L_{A_{eq},T}$ is the equivalent steady level of a fluctuating noise measured in dBA for a time T;[1] T is chosen according to the application, so, for example, for a school it could be the occupancy period. The BREEAM evaluation procedure for new offices

Table C.1 Representative sound pressure levels

Condition	*Sound pressure level (dBA)*
Threshold of hearing	10
Broadcasting studio	10–20
Living room in a quiet area at 7 a.m.	30
Typical business office	50–60
Listening to Chopin in a living room	50–65
Normal speech	55–70
Inside a British Rail train	60–70
Busy streets in urban area, e.g. Cambridge	70–75
Pop group at 20 m	100–110
Helicopter at 30 m	100–110
Threshold of pain	130

awards one credit if noise levels in large offices are equal to or below an $L_{A_{eq},T}$ of 45 dBA.[2]

It is also very common to see sound pressure level recommendations given as noise rating (NR) values. The approximate relationship between dBA and NR is:[3]

$$\text{dBA} = \text{NR} + 6.$$

Thus, if a recommendation for an air-conditioned conference room is NR25, the dBA equivalent would be about 31.

Fairly elaborate calculations are required to determine the sound pressure level in spaces – References 4 and 5 give typical procedures. A very approximate relationship[6] is:

$$\text{Noise level inside (dBA)} = \text{Noise level outside (dBA)} - \text{Average insulation (dB)}$$

Thus, if the noise level from a road is about 65 dBA and one is trying to achieve 30 dBA inside, an average insulation value of 35 dB would suffice.

References

1. Anon. (1987) Sound insulation and noise reduction for buildings. BS 8233:1987. British Standards Institution, London.
2. Prior, J. (ed.) (1993) *BREEAM/New Offices Version 1/93*, BRE, Garston.
3. Anon. (1986) *CIBSE Guide A1: Environmental Criteria for Design*, CIBSE, London.
4. Anon. (1993) *ASHRAE Handbook – Fundamentals*. Chapter 7: Sound vibration. ASHRAE, Atlanta.
5. Anon. (1986) *CIBSE Guide B12: Sound Control*, CIBSE, London.
6. Anon. (1988) Insulation against external noise. BRE Digest 338. BRE, Garston.

ILLUSTRATION ACKNOWLEDGEMENTS

The author and publishers would like to thank the following individuals and organizations for permission to reproduce material. We have made every effort to contact and acknowledge copyright holders, but if any errors have been made we would be happy to correct them at a later printing.

Individuals and organizations

Architects Design Partnership: 9.24
ASHRAE, Atlanta: 7.4
John Barnard: 5.2
Richard Bond: 15.1
Paddy Boyle: 9.20
The British Wind Energy Association, London: 7.5
Martin Charles: 11.1, 12.1
Chartered Institution of Building Services Engineers, London: A.3, A.5[*]
Cliché A. Attemand, L'Atelier du Regard, Orsay: 5.4
Combined Power Systems Ltd, Manchester: 7.2a
Peter Cook: cover photograph, 13.1, 13.6, 13.7
Edward Cullinan Architects: 8.3, 11.4
Electricity Association Services Ltd, London: 9.22
Max Fordham & Partners: 1.3, 4.5, 12.2, 13.4, 15.2, 15.6, B.1
Dennis Gilbert: 14.1
Short Ford & Associates: 13.2
Subinco: 4.6
United Distillers, Edinburgh: 9.13

Publications

Banham, R. (1969) *The Architecture of the Well-Tempered Environment*, The Architectural Press, London: 9.10
Clifton-Taylor, A. (1972) *The Pattern of English Building*, Faber & Faber, London: 1.2
Richards, J.M. (1958) *The Functional Tradition in Early Industrial Buildings*, The Architectural Press, London: 9.11, 9.12

[*]Pads of charts are available from CIBSE, 222 Balham High Road, London SW12 9BS

Index

Page numbers appearing in **bold** refer to illustrations and those in *italic* refer to tables.